Systematic Innovation
An Introduction to TRIZ $\left(\begin{array}{c}\textit{Theory of Inventive}\\ \textit{Problem Solving}\end{array}\right)$

John Terninko
Alla Zusman
Boris Zlotin

S^t_L

St. Lucie Press
Boca Raton London New York Washington, D.C.

Library of Congress Cataloging-in-Publication Data

Terninko, John
 Systematic innovation: an introduction to TRIZ / John Terninko, Alla Zussman, and Boris Zlotin.
   p. cm.
 Includes bibliographical references and index.
 ISBN 1-57444-111-6 (alk. paper)
   1. Problem solving — Methodology. 2. Creative thinking. 3. Creative ability in busines. 4, Technikcal innolvations. I. Zussman, Alla. II. Zlotin, Boris. III. Title.
 HD30.29.T47   1998
 658.5′75—dc21
                                                                       97-4687
                                                                         CIP

This book contains information obtained from authentic and highly regarded sources. Reprinted material is quoted with permission, and sources are indicated. A wide variety of references are listed. Reasonable efforts have been made to publish reliable data and information, but the author and the publisher cannot assume responsibility for the validity of all materials or for the consequences of their use.

Neither this book nor any part may be reproduced or transmitted in any form or by any means, electronic or mechanical, including photocopying, microfilming, and recording, or by any information storage or retrieval system, without prior permission in writing from the publisher.

The consent of CRC Press LLC does not extend to copying for general distribution, for promotion, for creating new works, or for resale. Specific permission must be obtained in writing from CRC Press LLC for such copying.

Direct all inquiries to CRC Press LLC, 2000 Corporate Blvd., N.W., Boca Raton, FL 33431.

Trademark Notice: Product or corporate names may be trademarks or registered trademarks, and are used only for identification and explanation, without intent to infringe.

© 1998 by CRC Press LLC
St. Lucie Press is an imprint of CRC Press LLC

No claim to original U.S. Government works
International Standard Book Number 1-57444-111-6
Library of Congress Card Number   97-4687
Printed in the United States of America   2 3 4 5 6 7 8 9 0
Printed on acid-free paper

Dedication

To Genrich Altshuller,
the father of the Theory of Inventive Problem Solving (TRIZ)

In recognition of many celebrations in the anniversary year 1996:

- The 70th birthday of TRIZ creator Genrich Altshuller (1926)
- The 50th anniversary of the inception of TRIZ (1946)
- The 40th anniversary of the first publication on TRIZ (1956)
- The 25th anniversary of Altshuller's last version of the Contradiction Table (1971)
- The 11th anniversary of Altshuller's last version of the Algorithm of Inventive Problem Solving (ARIZ) (1985)
- The 11th anniversary of the System of Standard Solutions (1985)

The Authors

John Terninko
For 16 years, Dr. John Terninko has been using and teaching QFD and Taguchi's philosophy as a consultant to corporations in North America, Central America, and Europe. For two years, he has been teaching and using TRIZ in North America. He has worked in the aerospace industry, the auto industry, and with durable and consumable products, and service organizations. John, who has written six books and presented 30 papers, is a principal with Responsible Management Inc., a close associate of GOAL/QPC, a founder of the QFD Institute, and a founder of the QFD Network.

Alla Zusman
Alla Zusman is the TRIZ Product Development Manager for Ideation International Inc., and has developed teaching materials, taught TRIZ and Ideation methodology, and developed the theoretical base for software products and publications. She has more than 16 years of TRIZ experience and coauthored seven books on TRIZ, including one with Genrich Altshuller, TRIZ founder. Alla has more than 3,000 hours of TRIZ teaching experience with more than 3,000 students, including more than 400 in the U.S. Her advancement of TRIZ includes a U.S. adaptation of the methodology

and development of new applications for solving scientific and business problems. She also developed the theoretical base for TRIZ software products.

Boris Zlotin

Boris Zlotin is Chief Scientist and Vice President of Ideation International Inc. where he organizes and provides analytical services while leading research and development work and publishing. With more than 22 years experience in TRIZ and engineering, he has coauthored nine books on TRIZ, including three with G. Altshuller, TRIZ founder. Boris has more than 8,000 hours of TRIZ teaching experience (more than 5,000 students), more than 6,000 hours of TRIZ consultation, and has facilitated the solving of more than 4,000 technological and business problems. His advancements of the methodology include Patterns of Evolution in different areas, U.S. adaptation of the methodology, and development of new applications in the solving of scientific and business problems. He developed the theoretical base for TRIZ software products.

Acknowledgments

There are two expressions whirling in my head as I think about how best to thank the people who have helped in the development of this material: "Those who cannot, teach — and those who can, do" and "The best way to learn something is to teach it." I believe I have experienced a third one: "The best way to show the world you know something is to write about it in an understandable manner." This work has acted as a kind of final exam, given to me by Alla Zusman, my teacher, mentor, friend, colleague and coauthor, and Boris Zlotin, my friend, mentor, teacher, colleague and spellbinder.

About five years ago I heard about something called the "Theory of Inventive Problem Solving (TRIZ)." Busy developing a consulting business in QFD and Taguchi, I didn't think it could get any better than Stewart Pugh's concept selection. But, thanks to Glenn Mazur, a long time friend and colleague, three years ago I met Zion Bar-El, Alla Zusman, and Boris Zlotin at a one-day TRIZ overview presentation. Since then, I've been caught in the whirlwind of learning and developing a new way of viewing the world.

I gratefully acknowledge Zion's ability to find and share opportunities at every juncture.

To Boris go my thanks for lending the full extent of his knowledge. His willingness to devote many days to my education and to give me the honor of being the second certified TRIZ missionary are deeply appreciated. Boris, who sees an application for TRIZ in every step of life, pushes the theory further with every breath.

Dana Clarke, the first certified TRIZ missionary, lent his additional year of TRIZ experience to help me through the rough spots. I am happy to recognize both his ability and kindness.

To Alla, who imbued both my understanding and this book with her logical, structured approach to TRIZ, many thanks.

I also would like to acknowledge the clients and course participants who offered suggestions which have been incorporated in this book.

Finally, thanks to the team of Maggie Rogers and Mary Ann Kahl. Maggie, with her nontechnical view of life, helped me get the point across and deserves the hearty acknowledgment of us all. Mary Ann, with a more technical background, tempered my rambling muse.

I must not forget Candy who had to live with my strange hours while I worked on this book — and suggested I not write another one this year.

John Terninko

Preface

There are many rituals practiced to help the subconscious break out of traditional thinking patterns. Accepted methods of encouraging group creativity come in the form of brainstorming and Synectics, with some other variations. Individuals have been known to work until exhausted, and then go to sleep expecting (or hoping) to wake up early with an inspiration.

Several books have been written on improving creativity. They offer ways to challenge the old paradigms or reduce mental blocks based on the idea that, "Knowledge is the stuff from which new ideas are made. Nonetheless, knowledge alone won't make a person creative."[1]

Dr. Albert Szent Györgyi, a Nobel laureate, once observed, "Discovery consists of looking at the same thing as everyone else and thinking something different." A key to innovation is making an old idea a new idea. But, in looking for the right solution to a problem, it is important to remember the philosophy of Emiè Chartier that, "nothing is more dangerous than an idea when it is the only one you have."[2]

The French psychologist Antwan Ribaut was interested in creativity and invention and looked for a common thread in the creative process. In a 1912 study, he demonstrated that creativity peaked at age 18 and then decreased for the rest of a person's life.[3] This is very discouraging for organizations whose employees are all more than 18 years old!

Years later Genrich Altshuller, creator of the Theory of Inventive Problem Solving (TRIZ), replicated Ribaut's study and found even more distressing news. Notice in Figure 1 that Altshuller's peak at 14 was less creative than Ribaut's.[4] In Boris Zlotin's 1980 study, it was hypothesized that this was the result of children's overloading with information and diminished belief in fairy tales, which are full of inherent contradictions. Also in Zlotin's study, creativity dropped quickly to a lifetime low at about age 21. The implication

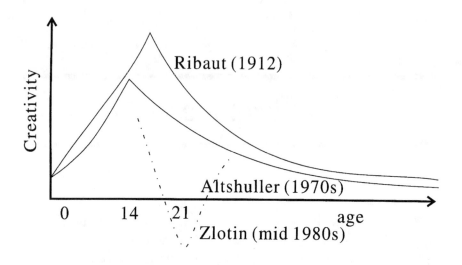

Figure 1. A comparison of the creativity curves of Ribaut, Altshuller, and Zlotin.

is that colleges and universities reduce creativity. This makes us wonder how many students can remember a teacher who encouraged original ideas. There was a wonderful movie in which a grandfather started college with his grandson. In their Economics class, the young instructor was lecturing on the Great Depression in the United States. The grandfather raised his hand and disagreed with the professor. He said that he had lived through the depression and it was not like the lecture. The grandfather was told he was the student and not the professor and that the professor was correct. How much of our lives are variations on that experience?

Though it appears to be in the same class as "jumbo shrimp," systematic innovation is not an oxymoron. Possibly because TRIZ specializes in eliminating contradictions, systematic innovation is not only possible, but highly effective. TRIZ offers examples which lead in new directions, turning psychological inertia into a whirlwind of new ideas — it gives a gentle smack on the forehead, often creating more alternative ideas than you care to consider.

For many, TRIZ is a new way of thinking which takes time to absorb. Each of the following chapters should provoke a breakthrough in your thinking patterns. Take your time to learn comfortably over an extended period of time. Learning a new language requires practice. As growing proficiency in a language shades our perception of the world, so will learning and practicing the methods presented in this book change your view of life. All around you will be opportunities for improvement, and TRIZ will offer you suggestions on how the improvements can be made. Feel free to leave off from the text, use what you learned, and pick up at a later date when ready for the next level. And be encouraged by the studies which have shown that creativity scores can increase over time.

The following table, developed by Boris Zlotin, illustrates the thinking patterns of adult, child and TRIZ-trained minds. This book aims at being a TRIZ introduction which will help you get to the thinking in column three.

Adult's Thinking	Children's Thinking	TRIZ Thinking
Fears contradictions and desires to avoid them.	Not sensitive to contradictions and lacks the desire to avoid them in arguments.	Loves contradictions and searches for them in problems. Understands that revelation and formulation of an obvious contradiction is a step toward its resolution.
Takes a metaphysical approach, considering objects, processes and phenomena separately rather than systemically.	Takes a syncretistic approach and desires to connect "everything with everything."	Takes a systematic approach and desires to find the connections between remote objects, processes and phenomena which often look as though they are not connected.
Uses an unorganized combination of various types of deductions, often applied erroneously.	Uses traduction, a type of deduction, erroneous from the viewpoint of classical logic, where inferences are made from one specific fact to another specific fact.	Uses deduction by analogue. Shows transition of deductions, ideas and solutions between systems (an organized combination of induction, deduction and traduction).
Follows a combination of logical thinking and natural intuition.	Follows an innate ability to produce an intuitive deduction.	Follows a combination of logical thinking with purposely formed intuition.
Has "law obedience," by which laws are intuitively known or verbalized.	Has "creation of laws," with a spontaneous search and development of intuitive and verbalized laws.	Has a purposeful search and development of laws and the verbalization of the intuitive laws.
Attempts to brainstorm the difficult problem with one shot, giving up the solution in the case of failure.	Attempts substitution of the problem. If unable to solve a problem, purposely modify the conditions and rules to eliminate it.	Attempts substitution of the problem with another one which can be solved by known rules.

Children are creative, but do not have the necessary knowledge to invent. Adults have knowledge, but have lost their natural creativity. Truth can help to acquire knowledge without losing creative imagination.

Contents

1 Introduction to TRIZ, the Theory of Inventive Problem Solving 1
2 Innovation Situation Questionnaire, ISQ 29
3 Problem Formulation 47
4 Contradiction 65
5 The Ideal Design 91
6 System Modeling, Substance-Field Analysis 113
7 Patterns of Evolution 127
8 Implementation 147

Appendix:
- A Physical Effects and Phenomena 155
- B 39 Parameters — Definitions 161
- C 40 Principles — Definitions and Examples 165
- D 40 Principles — Frequency of Use 177
- E Contradiction Table 179
- F Containment Ring Case 187
- G Resources 195

Glossary 197

References 199

Bibliography 203

Index 205

1 Introduction to TRIZ, the Theory of Inventive Problem Solving

Synergy

Three powerful tools are available to enhance an organization's design process. First, **Quality Function Deployment (QFD)** translates all relevant customer information into the language of the engineer to facilitate product design. Originally, Pugh's concept selection and brainstorming were the main approaches to innovation.[5] By adding the **Theory of Inventive Problem Solving (TRIZ)** to QFD, quantum improvements in the areas of innovative product and process design are realized. However, TRIZ assumes the given problem is appropriate and that it does not do the final design work. A final tool, **Genichi Taguchi's Philosophy of Robust Design**,[6] will identify component values which enable a design to perform on target independently of uncontrolled influences.

QFD + TRIZ + Taguchi = Customer-Driven Robust Innovation

This book offers an overview of one element of customer-driven robust innovation, TRIZ.

The effect of these three tools on an organization's ability to attain its objectives is shown in Figure 1. Filled-in circles show that a column has significant impact on a row. The figure illustrates the significant impact of TRIZ on product innovation. Quarter-circles show a moderate impact and empty circles show a weak impact. The roof illustrates the significant positive synergy between QFD, TRIZ and Taguchi.

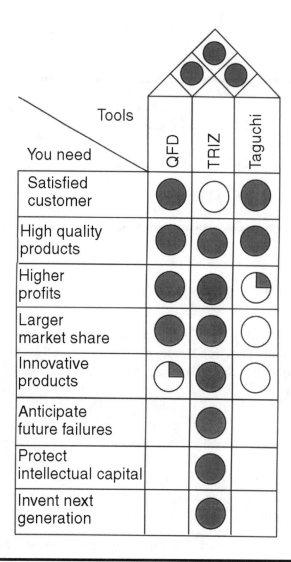

Figure 1. The relationship of QFD, TRIZ, and Taguchi.

Abstract

The word *systematic* conjures up the image of sequential activities that can be performed repeatedly to yield a desired result. *Innovation* has an association with creativity, which frequently implies unpredictable and erratic processes. Despite this, the term "systematic innovation" is not an oxymoron.

The pillar of the Theory of Inventive Problem Solving (TRIZ) is the realization that contradictions can be methodically resolved through the application of innovative solutions. This is one of three premises upon which the theory is built: (1) the ideal design is a goal, (2) contradictions help solve problems, and (3) the innovative process can be structured systematically.

The assumption that we cannot harness and control innovative processes is not only limiting, but also incorrect. Inspiration need not be random. TRIZ practitioners continually demonstrate that applying common solutions for the resolution of contradictions, identified as effective when applied to parallel problems in the world patent base, radically improves the design of systems and products. Once a problem is structured as a contradiction, methods exist for resolving that contradiction. These methods are rapidly evolving and are more widely available. The key to understanding why systematic innovation is possible is understanding the common solutions among the innovative world patents.

Altshuller questioned our common assumptions about the creative process. In his book *Creativity as an Exact Science*, Genrich Altshuller quotes the playwright Rosoy on the subject:

> Everyone knows that the act of creativity is not arbitrary. It does not respond even to the mightiest effort of will or peremptory command. It seems to me that the artist thinks right up to the moment of creation and also afterwards, but at the time of the act of creation itself does not consciously reflect on it.[7]

Rosoy articulates this problem that plagues inventors and artists: How can we control processes that seem to elude our grasp?

Brainstorming is a popular technique for capturing the ideas that lurk in our subconscious. Frequently, brainstorming methods work by trying to make us think about a problem from a new perspective. Questions typical of this approach are:

- Why are you working on this problem?
- What if you wanted the opposite effect/solution?
- What if the process could be performed faster?
- What happens if you change a substance?

An attempt to improve the method of brainstorming (called *Synetics*) was developed by William Gordon in the 1960s.[8] Gordon found that synergy exists between brainstorming, critical thinking and analogous thinking. A framework can be created to connect these mental processes in order to capture this synergy. One technique is to pretend that you are the product being designed and then explain how to solve the problem "from the inside." A similar approach considers the design challenge as the plot of a fairy tale or science fiction story — you simply explain how the hero or heroine resolves the conflict.

All of these methods, except Synectics, have a short learning curve and are more effective than using random trial and error to solve design problems. In contrast, Thomas Edison was a proponent of the trial and error method, believing that genius is "1% inspiration and 99% perspiration."[9] It is worth noting, however, that Edison had several hundred assistants and could afford many thousands of trials for an idea…using this method, Edison often needed as many as 50,000 trials for an invention.

Traditional processes for increasing creativity have a major flaw in that their usefulness decreases as the complexity of the problem increases. At some point, the trial and error method is used in every process, and the number of necessary trials increases with the difficulty of the inventive problem. Some solutions may even require more than one generation of problem solvers. It was Altshuller's quest to facilitate the resolution of difficult inventive problems and pass the process for this facilitation on to other people. His determination to improve the inventive process led to the creation of TRIZ.

In working toward his goal of developing the "science" of creativity, Altshuller's central questions were:

- How can the time required to invent be reduced?
- How can a process be structured to enhance breakthrough thinking?

In trying to answer these questions, Altshuller realized how difficult it is for scientists to think outside their fields of reference, because that involves thinking with a different technology or "language." Altshuller (and his growing number of colleagues) began to analyze innovative solutions to existing problems as they were identified in the world patent base. By identifying existing patterns of creative problem solving across technologies, the problem of narrowed vision due to specialization is overcome and the inventive process is dramatically improved. Now anyone who can think can be inventive…but also, talented inventors can become more effective. By the 1980s, Altshuller estimated that close to 100 TRIZ institutes and "schools of thought" had been established. These organizations expanded the boundaries of the TRIZ theory by successfully applying their insights on problem solving to real situations, and then identifying parallel innovative principles from the application of the processes that shaped their solutions.

History of the Theory of TRIZ

In 400 A.D., the Greek Pappos defined the word *heuristics* as the science of making discoveries and inventions. The words *discovery* and *invention* themselves had a broader meaning, including the productive work of artists, politicians, generals, etc., but heuristics identified a process for problem solving. Altshuller initiated a renaissance in heuristics through his process for systematic innovation.

TRIZ is the acronym for the Russian words:

> Теория Решения Изобретательских Задач

It is pronounced "treez," with a long "e" and a rolled "r." The translation "Theory of Inventive Problem Solving" produces the acronym TIPS. Altshuller's *Creativity as an Exact Science* offers the translation "Theory of the Solution of Inventive Problems (TSIP)." Additionally, some organizations refer to TRIZ as Systematic Innovation. A search on the World Wide Web for information about TRIZ and its application should use all four names. Whatever its designation, TRIZ represents a unique means of increasing innovation and radically improving design. This powerful tool eliminates the need for compromise and tradeoff caused by conflict between different performance measures. Instead, TRIZ celebrates the identification of conflict as the opportunity for improvement and, in so doing, refines the design process.

The Face Behind the Revolution

Genrich Altshuller (1926), born in the former Soviet Union almost 10 years after the beginning of the Bolshevik Revolution, showed a penchant for creative innovation early in life. As a young boy he liked to swim under water. When he was only 14, he invented, developed and tested a device that generated oxygen from hydrogen peroxide. Using this device, he could remain under the surface for longer intervals. Without realizing it, Altshuller was already applying the concept of solving design contradictions without compromise by using readily available resources.

To accumulate oxygen, the gas should be pressurized or converted into liquid. Both require complex equipment, which was not available. Altshuller's solution used a phase transformation of the compound liquid hydrogen peroxide to gases, one of which is oxygen. At 16, Altshuller received his first author's certificate for this underwater device.

Altshuller continued his inventive career under Josef Stalin's regime. At that time in the Soviet Union, there were two types of registrations for inventions. A patent was rarely given, so most inventors applied for an author's certificate. The Soviet government owned the intellectual property that the author certificate documented, so the certificate was merely an acknowledgment of the inventor's contribution. Ironically, it is the simple, direct format of the author certificate that facilitated Altshuller's research into the inventive process. The government's collection of author's certificates provided him with a huge database of valuable information. During the formulation of TRIZ, Altshuller and colleagues reviewed tens of thousands of author's certificates and patents (Figure 2).

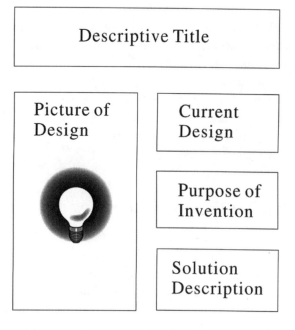

Figure 2. A mockup of a Russian author's certificate.

During World War II, Altshuller joined the Army. After the war, he was assigned to the Navy as an inspector of inventing, a job which he saw as an opportunity to help inventors find creative solutions to technical problems. At first, Altshuller turned to psychology for insights into unlocking the inventors' creativity. Ultimately, these investigations into the behavioral sciences were not as productive as his observation of design patterns that began to emerge in his work-related review of thousands of author's certificates.

In 1946, Altshuller decided that he must create a new science for the theory of invention. His achievements during this period were staggering. Within two years he established a foundation for TRIZ by studying thousands of author certificates. The multipage documents included a cover sheet, a one-page sketch and a short invention description. This simple format made it easy to identify underlying patterns of the inventive process.

Altshuller identified patterns frequently used in the more innovative patents. He defined five levels of invention based upon the criteria: how remote the knowledge used for the solution concept was from the inventor's field, the theoretical number of trials to a solution, and how substantial the change was from the original design to the solution.

Patents representing a simple modification to a design were assigned to the lowest level. Patents that changed the system in some way were considered more inventive, while patents introducing a new science were considered the most innovative. These innovative patents provided solutions to contradictions, and these solutions often represented identifiable points along repeatable lines of evolution. Altshuller's research replaced the unpredictable "Eureka!" of the stereotypic mad scientist or absent-minded professor with specific patterns of design evolution which could be followed by the real-life problem solver (Table 1).

These patterns identified in the development of a design contain two major components: regularities in design evolution, and principles used in innovative solutions. Altshuller's observations led to an additional breakthrough; since the evolution of engineering design was a process governed by definable laws, it could be taught. Using their newfound knowledge, Altshuller and his colleagues created many effective civil and military inventions. A new way to collect spilled oil, a method to rescue the crew from a sunken submarine, an innovative environmental suit for rescuers entering a mine — these were just a few of the technologies developed. More important than each individual innovation, however, was the fact that a revolution in the field of inventive problem solving had begun.

Table 1. Genrich Altshuller (Altov)

First Invention at age 14.

1946–1948

Studied 200,000 patents.

Selected 40,000 patents as representing the most effective solutions.

Evolution of an engineering system is not a random event, but governed by certain patterns.

Inventiveness and creativity can be taught.

1985

Published 14 books and hundreds of papers.

Altshuller and his boyhood friend, Raphael Shapiro were excited about their progress. Though fearful of the possible consequences, they decided to voice their concern about the future of inventions in the Soviet Union in a 1948 letter to Stalin. They criticized the inventive process used throughout the nation and offered some measures to improve the methodology. Their proposed improvements were an embryonic form of TRIZ. Unfortunately, their patriotism and valuable ideas were not rewarded. Altshuller and Shapiro were charged with "inventing with the purpose to do harm to the country." After a year of interrogation and torture, they were sentenced to 25 years in a prison camp above the Arctic Circle.

What would have been a hellish existence for most people became a time of significant intellectual growth and productivity for Altshuller. The prison camp contained dozens of professors, eminent scientists, musicians, and artists, all of whom were jailed during Stalin's great Purges. As a result, Altshuller's education continued. Because fellow prisoners were happy to have someone who was eager to learn and listen for hours, the prison camp became Altshuller's private university. The worst punishment for Altshuller was the prohibition on writing. A prisoner could be beaten cruelly and placed in a cell if he were found in possession of a notebook. Despite this considerable obstacle, Altshuller continued to develop the science of innovation. With his remarkable memory and analytic ability, Altshuller used this time to work out laws for technical system development and methods of solving inventive problems.

Stalin died in 1953, and Altshuller and Shapiro were released one year later. The two men continued to develop TRIZ, publishing their first article on principles of their theory in a 1956 issue of a scientific magazine entitled

The Questions of Psychology.[10] Shapiro wrote the first book on TRIZ, but subsequently lost interest in furthering the theory.[11]

During the next decades, Altshuller's results attracted professionals from many disciplines who adopted, adapted and expanded his methodology. In 1974, Boris Zlotin became an active proponent, and Alla Zusman joined the TRIZ group of interested professionals in 1981. Applications in the real world advanced and verified the theory. Methods for solving problems were improved. TRIZ methodology was applied to problems in science, business, management and other areas. As TRIZ became a vehicle for creative education, it also became the subject of newspaper and magazine articles, books, and a regularly scheduled TV show for children.

Yet even after Stalin's death, a constant fight with the State Committee of Inventive Affairs and the Society of Inventors interfered with the use of TRIZ. The Soviet government was unwilling to listen to an intellectual Jew. In such a restrictive climate, it was difficult for even the most influential people working in industry to take advantage of these inventive methods. Altshuller's difficulties were compounded by the fact that he and his colleagues were not able to use these techniques for financial gain. Under the pseudonym Altov, Altshuller wrote science fiction stories to earn his living. But here again, he found an application for TRIZ in the creation of many of the ideas for his futuristic devices and creatures.

Despite these difficulties, a strong and loyal following continued to grow. During the 1970s, translations of Altshuller's books and articles circulated in Germany and Poland, eventually reaching Japan, the U.S.A. and other Western countries. Development of TRIZ in the Soviet Union accelerated after Gorbachev's Perestroika. Private companies were ready to use TRIZ for solving complicated technical management issues and other problems because they could see the advantage it gave them in a free-market economy.

By 1985, Altshuller had written over 14 books, including several with Boris Zlotin and one with Alla Zusman. Only two of Altshuller's books have been translated into English. Altshuller's key findings are explained in these books, which reflect his study of over 200,000 patents, focusing on 40,000 identified as containing the most innovative design solutions (Table 2).

When failing health restricted Altshuller's working hours in 1985, he began to concentrate on creativity in more general terms. It is the collection of techniques associated with Altshuller prior to his illness that has become known as classical TRIZ.

Considering this history, it is reasonable to ask, "How prevalent is TRIZ in the former Soviet Union?" Prior to 1986, there was little opportunity to

Table 2. Altshuller's Key Findings & Developments

Levels of Innovation
Contradictions
  Technical
    40 Inventive Principles (1956–1971)
    39 Engineering Parameters
  Physical
    Four Separation Principles (1979)
Ideality (1956)
76 Standard Solutions (1974–1985)
Patterns of Evolution (1969—1979)
ARIZ (Algorithm of Inventive Problem Solving) (1959–1985)
Substance–Field Analysis (1977)

---

apply TRIZ in real situations. Individuals who made an effort to implement the methodology faced many restrictions. However, a large number of individuals were trained during this time. But where the application of TRIZ would have been effective in business and industry, there was no means of establishing an organized network of practitioners. Today, TRIZ is used by financial institutions, as part of the curriculum in private schools, and in politics.

Pragmatic North Americans always are looking for ways to improve the innovative process, and many organizations work to develop internal expertise. Typically, a consultant provides some training for a particular methodology and demonstrates the process by creating concepts for solving a real problem. The organization begins internal training in order to solve innovative problems without the consultant. The combination of hiring a consultant for the challenging problem, then using the difficult problem as a training case, efficiently develops an organization's experience base. Anxious to apply their knowledge to real problems, TRIZ experts coming from eastern Europe found a receptive audience in the United States.

Rockwell Automotive used a TRIZ consultant to reduce the number of components in a brake design from 12 to four. These improvements reduced the cost of the brakes by 50 percent.[12] Ford Motor Co. was looking for a solution to a transaxle bearing which would gradually move away from the correct position under heavy loads. The TRIZ application generated 28 design concepts. One of the more interesting concepts featured a bearing with a smaller coefficient of expansion, which took advantage of the higher temperatures generated during increased loads — the heavier the load, the tighter the bearing.

Overview of the Following Chapters

The key elements of classical TRIZ, plus two later developments that are presented in this book. The tools for applying this methodology can be broken into two groups: analytic tools for structuring the problem and knowledge-based tools for providing the database from which concepts are generated. During the development of TRIZ, Altshuller worked on effective problem solution, assuming the identified problem was appropriate. In actual application, a way of clearly identifying the problem was needed. Two additional analytic tools for defining the problem environment and formulating the problem have been developed for use. Chapters 2 and 3 describe these new tools. The classical TRIZ tools are introduced in Chapters 4 through 7. Chapter 8 presents implementation considerations (Table 3).

Table 3. Layout of the Book

Chapter 1.	Introduction to The Theory of Inventive Problem Solving, TRIZ
Chapter 2.	Innovation Situation Questionnaire, ISQ
Chapter 3.	Problem Formulation
Chapter 4.	Contradiction
Chapter 5.	The Ideal Design, Ideality
Chapter 6.	System Modeling, Substance–Field Analysis
Chapter 7.	Patterns of Evolution
Chapter 8.	Implementation

TRIZ practitioners estimate that more than two million patents worldwide have now been reviewed to identify patterns and regularities contributing to further the refinement of TRIZ. The different schools for TRIZ and individual practitioners continue to improve the methodology. Zlotin, Zusman and their team at the Kishnev TRIZ School in Moldova began restructuring and improving TRIZ in 1985[13] (Table 4). Identifying weaknesses in the existing TRIZ methodology, Alla Zusman and Boris Zlotin created some new tools for applying TRIZ to real situations. These two experts brought their applications to the United States. Other TRIZ specialists have also immigrated to the west.

Table 4. The Growth of TRIZ

Boris Zlotin, Alla Zusman and associates have:
  Revised and restructured TRIZ Tools (1992)
  Expanded the Lines of Evolution to more than 250
  Enhanced ARIZ (1995)
  Created new tools
    Problem formulation process (1992)
    Systems of Operators (1993)
    Developed Applications:
      Anticipatory Failure Determination
      Directed Evolution
      Revealing new tasks for improvement

These new instruments, along with those developed in classical TRIZ, provide the user with a work box full of tools.[14] The master craftsman has experience with all the tools of his or her trade, and knows which is most appropriate for a given problem. Two philosophically different software packages exist today to reduce the time needed to solve innovative problems successfully. One has been developed by Valery Tsourikov of Invention Machine in Boston, MA. The other is by Ideation International, Southfield, MI, from which most of this book was developed. This text is software free; it will help you solve design/inventive problems independent of any software, and it complements either software package.

Degree of Inventiveness

Altshuller defined an inventive problem as one containing at least one contradiction. He defined a contradiction as a situation where an attempt to improve one feature of the system detracts from another feature. After initially reviewing 200,000 patent abstracts, Altshuller selected 40,000 as representative of inventive solutions. The remainder involved direct improvements easily recognized within the specialty of the system. Altshuller separated the patents' different degrees of inventiveness into five levels (Table 5). It should be noted that a problem cannot be ranked into one of the five levels — only the solution to a problem.

Table 5. Levels of Innovation

1. Apparent or Conventional Solution: 32%
    Solution by methods well known within specialty
2. Small Invention Inside Paradigm: 45%
    Improvement of an existing system, usually with some compromise
3. Substantial Invention Inside Technology: 18%
    Essential improvement of existing system
4. Invention Outside Technology: 4%
    New generation of design using science not technology
5. Discovery: 1%
    Major discovery and new science

Levels of Innovation

Between 1964 and 1974, the patents under review were evaluated twice to determine relative frequencies for the different levels of innovation. The percentages given here are from that period. More recent evaluations have not been done to check these numbers.

Level 1 represented 32% of the patent inventions and employs obvious solutions drawn from only a few clear options. Level 1 innovations are not inventions but narrow extensions or improvements of the existing system, which is not substantially changed. Usually a particular feature is enhanced or strengthened. Examples of Level 1 solutions include increasing the thickness of walls to allow for greater insulation in homes or increasing the distance between the front skis on a snowmobile for greater stability. These solutions may represent good engineering, but contradictions are not identified and resolved.

Level 2 solutions offer small improvements to an existing system by reducing a contradiction inherent in the system while still requiring obvious compromises. These solutions represented 45% of the inventions. The improvement is usually found through a few hundred trial and error attempts and requires knowledge of only a single field of technology. The existing system is slightly changed, and includes new features that lead to definite improvements. The new suspension system between the track drive and the frame of a snowmobile is a Level 2 invention. The use of adjustable steering columns to increase the range of body types that can comfortably drive an automobile is another example at this level.

Level 3 inventions, which significantly improve existing systems, represented 18% of the patents. At this level, an inventive contradiction is resolved within the existing system, often through the introduction of some entirely new element. This type of solution may involve several hundred ideas, tested by trial and error. Examples include replacing the standard transmission of a car with an automatic transmission, or placing a clutch drive on an electric drill. These inventions usually involve technology integral to other industries, but not widely known within the industry in which the inventive problem arose. The resulting solution causes a paradigm shift within the industry. A Level 3 innovation is found outside an industry's range of accepted ideas and principles.

Level 4 solutions are found in science, not in technology. Such breakthroughs represented about 4% of inventions. Tens of thousands of random trials are usually required for these solutions. The solutions lie outside a technology's normal paradigm and involve using a completely different principle for the primary function. In Level 4 solutions, the contradiction is eliminated because its existence is impossible within the new system. That is, Level 4 breakthroughs use physical effects and phenomena that had previously been little known within the area. A simple example involves using materials with thermal memory (shape-memory metals) for a key ring. Instead of taking a key on or off a steel ring by forcing the ring open, the ring is placed in hot water. The metal's memory causes it to open for easy placement of the key. At room temperature, the ring closes. Another example is cleaning surfaces or burrs using cavitation created by ultrasound. Microlevel explosions occur on the surface to be cleaned through the use of directed ultrasound.

Level 5 solutions exist outside the confines of contemporary scientific knowledge. Such pioneering work represented less than 1% of inventions. These discoveries require lifetimes of dedication, for they involve the investigation of tens of thousands of ideas. This type of solution occurs when a new phenomenon is discovered and applied to the inventive problem. Level 5 solutions, such as lasers and transistors, create new systems and industries. Once a Level 5 discovery becomes known, subsequent applications or inventions occur at one of the four lower levels. For example, the laser, technological wonder of the 1960s, is now used routinely as a lecturer's pointer and a land surveyor's measuring instrument. In 1995, a home owner could buy a laser device the size of a bar of soap to measure distances up to 7 meters. This evolution within the laser industry illustrates that a solution's level of inventiveness is time-dependent.

Altshuller focused his investigation on the principles used in Level 2, 3 and 4 solutions. Level 1 solutions mostly were ignored because they need not be innovative. Because Level 5 solutions require understanding a new natural phenomenon and therefore offered a very small sample, there were no recognizable parallels among these inventions. But Altshuller believed that he could help anyone develop Level 2, 3 and 4 innovations. His desire to develop this science of invention inspired the study of hundreds of thousands of patents. Through this effort, Altshuller initially identified 40 principles used consistently to resolve technical contradictions.

The enthusiastic inventor needs to understand the discrepancy between the pace of social change and the faster pace of technological progress. The difference between these two timelines complicates the success of upper level innovations. In *And Suddenly the Inventor Appeared*, Altshuller writes, "If one chooses to develop a completely new technical system when the old one is not exhausted in its development, the road to success and acceptance by society is very harsh and long. A task that is far ahead of its time is not easy to solve. And the most difficult task is to prove that a new system is possible and necessary."[15] In other words, the inventor must be cautious because designs that are too advanced may not be accepted by the public and support for further development will be limited. Introducing several incremental improvements to an existing system is a better strategy. The initial reaction to radar when it was introduced during World War II illustrates a common response to new inventions. Radar radically increased a submarine crew's awareness of approaching aircraft. However, the submarine captains of one country refused to use the newly installed devices. Since they were now aware of so many more aircraft, the captains thought the radar must be attracting the planes. Today their reaction may seem strange, but it exemplifies peoples' initial resistance to technological innovation. For a more contemporary example, consider that computer programmers initially saw no reason to provide a monitor with a computer. In this example, there are two issues the designers did not understand: they did not anticipate the needs of the customer, nor did they realize the usefulness of the CRT for themselves.

Inventions involving Levels 1, 2 and 3 are usually transferable from one discipline to another. This means that 95% of the inventive problems in any particular field have already been solved in some other field.

The majority of patents fall within four major technologies: mechanical, electromagnetic, chemical and thermodynamics. Think of how much knowledge an individual has, then think of how much more knowledge is recognized collectively by a corporation, an industry, a society and in the universe

(Figure 3). If the investigator's knowledge base contains all information within an industry, then fewer solutions are available than if the investigator knows everything there is to know in society.

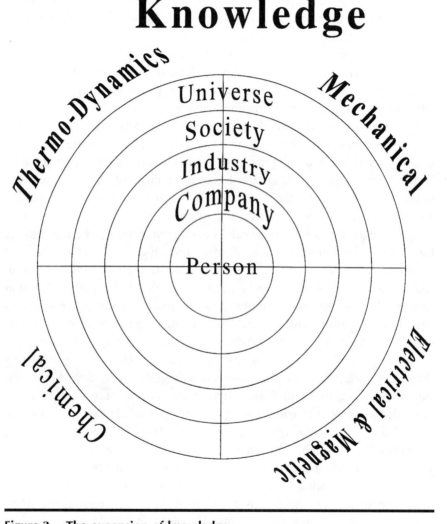

Figure 3. The expansion of knowledge.

The level of inventiveness within a field can be categorized as having a range from Level 1 (personal knowledge) to Level 5 (universal knowledge). Universal knowledge representing the composite of all known information.

A paradigm shift for a mechanical engineer who is working on a problem can be provided by his or her child, using new-found knowledge from a chemistry course. Similarly, the level of innovation increases when inventors move outside of their normal fields of inquiry.

A method that redefines a problem in order to move it to a lower level, relative to the solutions position in the pattern of design evolution, will facilitate the creation of better designs in less time. A higher level problem made into a Level 1 problem will often contain recommendations foreign to the problem solver or beyond his or her comprehension. For example, if a recommendation to reduce the effect of centrifugal force is dependent upon a liquid pressing against the axis of rotation, rather than to the walls, the problem solver might think the concept came from another planet. Few technical people are aware of the Weissenberg phenomenon for material with a particular internal structure. This phenomenon is not taught in most physics courses. When interviewed for this book, a 50-year-old senior physics professor said he was unaware of any liquids that clung to the axis of rotation instead of being thrown outward by centrifugal force. Nevertheless, anyone who uses a bread machine can observe the dough climbing up the axis of rotation because yeast's cell structure elongates in the direction of the axis of rotation. A reference identifying all special effects of physics and chemistry would be useful. Presently, the only references in English are contained within software.

Inventive problems are often mistakenly considered to be the same as engineering, technological and design problems. However, the inventor is looking for ways to solve problems *by eliminating contradictions*. This is the crucial difference between their work and the efforts of others involved in the design process. Once the concept or inventive solution is found, it is necessary to use the skills of engineers, technologists and designers. Their work is what transforms the concept into the finished product wanted by the public.

Another influence that slows the development of innovation is psychological inertia. Researchers have their favorite directions for investigation, and these generally look like a line vector located within or near their specialties. Since all professionals have some knowledge in other disciplines, the line vector is more accurately a cone that encompasses pieces of other disciplines in the accepted search region. Psychological inertia influences researchers to move in the same direction as they have on successful project searches in the past. This situation resembles a laboratory rat that traces only one path in the maze of world knowledge.

Psychological inertia also recalls the joke about a person looking for his or her car keys under a street light. A stranger offers to help and after several minutes asks where the keys were lost. The car's owner responds, "They were dropped in the shadows by the car, but the light is better here." If you look in the wrong place, innovative concepts will not be found. Processes that encourage creativity provide many branches radiating from one technological direction, but do not access solutions that exist in another technology. Of course, this depends upon team composition. Figure 4 demonstrates the futility of looking in the wrong direction: No matter how sophisticated a process you use, it will not find a concept outside the area of research.

Figure 4. A misdirected search.

Several search directions become possible using the brainstorming technique introduced by A. Osborne in the 1940s.[16] Recognizing that a problem solving team is composed of some idea generators and some idea critics, he created a structure using both skills. A block of time is provided for any idea to be presented. This time allows the subconscious ideas to penetrate the constraints of the more conservative conscious mind. Such an approach improves random searches, but it is only effective for generating Level 1 and 2 innovations.

The weakness of the creativity approaches is even more dramatic when different technologies are added to the feasibility space, as shown in Figure 5.

Creating an interdisciplinary team also poses difficulties. It is virtually impossible to have all the members know everything about some area and have all the relevant areas represented. Consider any of your past or present projects, then record the areas of expertise and depth of knowledge of the team's members. You will probably note that there are some relevant areas that are not covered by the team, and that the depth of the members' knowledge is limited.

Solving design problems would be simpler if one could locate and hire the appropriate solver or surrogate solver. This is equivalent to hiring a consultant. But how do you know which consultant to hire? That hiring process in itself involves identifying your problem's need, which is one of the early steps in finding a solution. The good news is that once you know the technical expertise required, you may not need the consultant — particularly if you used the problem formulation process covered in Chapter 3.

Genrich Altshuller was driven by the belief that the "creative potential of the inventor is increased" when more knowledge becomes available. Using Altshuller's fundamental principles increases the knowledge available to a designer. Initially, Altshuller summarized this information as the 40 principles used to resolve technical contradictions between pairs of 39 engineering parameters.

Transferable Solutions Within Patterns of Innovation

While searching the patent fund, Altshuller recognized that the same fundamental problem (contradiction) had been addressed by a number of inventions in different areas of technology. This was more apparent if the problem was presented in technology-free terminology. He also observed that the same fundamental solutions were used over and over again, and that the implementations were often separated by many years. If some means of accessing the applications of these fundamental solutions were available to inventors, the number

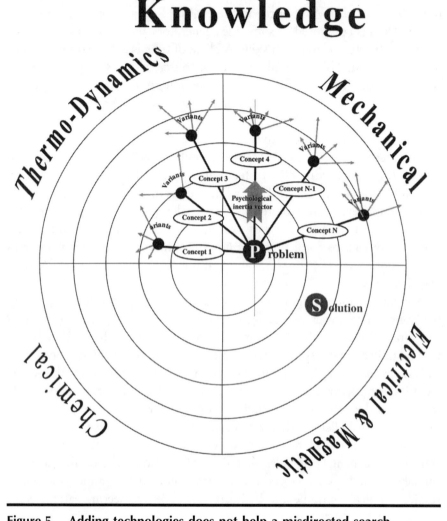

Figure 5. Adding technologies does not help a misdirected search.

of years between applications would decrease. Thus, the innovation process becomes more efficient, the time between improvements is reduced and the line separating different technologies is crossed more often. Let us explore several inventions that all used the same principle — a concept that was not initially considered by an artificial diamond manufacturer.

An organization producing artificial diamonds splits the crystals at the fracture to produce usable diamonds. Unfortunately, this process often results in new fractures. A process improvement team of engineers would not be inclined to look at agricultural patents for possible ideas. More importantly, the current world patent base is not structured for easy reference, though it would be useful to be able to look at the patents for devices that caused objects to come apart, fly apart or explode apart. Altshuller sorted patents in a manner that enhances innovation. By stripping away the technical subject matter, he found that the same problem was solved over and over again. Only a limited number of principles were necessary to explain the majority of inventions. If the diamond splitting team looked to agriculture, or had access to a database from Altshuller, they would find the following innovative solutions.

Invention 1. Sweet Pepper Canning Method

Before canning sweet peppers, the stalk and seeds must be separated from the pod. Before the invention described below, this was done manually. Automation was difficult because the pods were not uniform in shape or size.

This innovative method for canning sweet peppers begins with the pods being placed in an airtight container. The pressure is gradually increased to 8 atmospheres. The pods shrink and this results in fracturing at the weakest point, where the pod top joins the stalk. Compressed air penetrates the pepper at the fractures, and the pressure inside and outside the pepper equalizes. The pressure in the container is then quickly reduced. The pod bursts at its weakest point (further weakened by the fractures), and the top flies out along with the seeds (Figure 6).

Figure 6. Sweet pepper canning method.

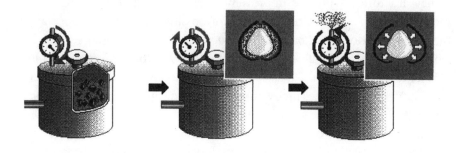

Figure 7. Shelling cedar nuts.

Invention 2. Shelling Cedar Nuts

The process for shelling cedar nuts is conceptually similar. The nuts are placed underwater in a pressure cooker. Heat is applied until the pressure reaches several atmospheres. The pressure is then quickly dropped to one atmosphere. After the overheated, high pressure water penetrates the nuts, the sudden pressure drop causes the shells to break and fly off (Figure 7).

A similar procedure is used for shelling krill — a small ocean crustacean.

Invention 3. Shelling Sunflower Seeds

Have you ever wondered how sunflower seeds are shelled? One method of shelling sunflower seeds involves loading them into a sealed container, increasing the pressure inside the container, and then having the encased pods flow out of the container. The pressure drops very quickly and the air that penetrated the husks under high pressure expands, thereby splitting the shells (Figure 8).

Invention 4. Producing Sugar Powder

A similar technique using lower pressures breaks sugar crystals into powder. Now, consider a nonagricultural application of the principle.

Invention 5. Filter Cleaning

A filter used to remove dust from air consists of a tube in which the walls are coated with a porous, felt-like material. When air passes through the tube,

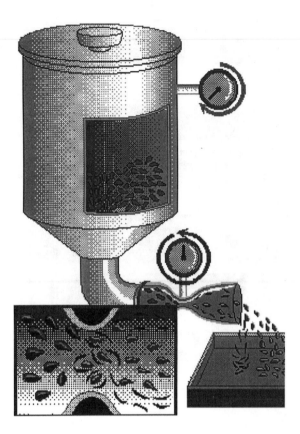

Figure 8. Husking sunflower seeds.

the dust particles are trapped in the pores. Cleaning such a filter is difficult. However, the filter can be cleaned by disconnecting it from the system, sealing it, exposing it to a pressure of 5 to 10 atmospheres, and then quickly dropping the pressure to 1 atmosphere. The sudden change in pressure forces the air out of the pores, along with the dust. The dust particles are carried to the surface of the filter where they are easily removed (Figure 9).

These five inventions occurred in two different industries at different times. If later inventors have knowledge of these earlier concepts/solutions, their tasks can be more straightforward. Unfortunately, interdisciplinary barriers make this information unavailable. The effort to solve the diamond splitting problem could have been reduced if the researchers had read any of the above patents. Note that the principle is the same, but the design of the system and its procedures are different. Concept creation is the task of the inventor and system design is the responsibility of the engineer.

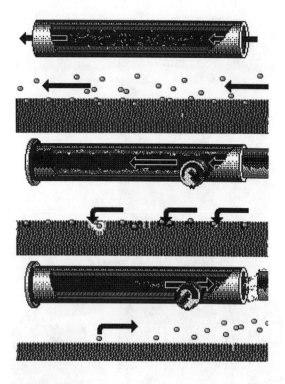

Figure 9. Filter cleaning.

Invention 6. Splitting Imperfect Crystals

The diamond manufacturer eventually received a patent for this solution. Crystals are placed in a thick-walled, airtight vessel. The pressure in the vessel is increased to several thousand atmospheres, then quickly returned to normal. This sudden change in pressure causes the air in the fractures to break the crystals.

Past and present applications of TRIZ use principles common to several disciplines. For example, Ford Motor Co. has used TRIZ several times, especially in the area of reliability testing. Mike Lynch, supervisor of reliability systems in Advanced Vehicle Technology, once said "You must be able to look past the fact that this is a bell pepper and not a bumper. That is not the issue. The issue is the pressure change and removal of rust off bumpers. As an engineer or scientist, you have to make that correlation."[17] Lynch's remark underscores how widening the knowledge base improves designs.

After seeing the possibilities of TRIZ, one engineer for an international organization said, "What is the big deal? All you have done is generate several creative ideas. Getting the idea is only a small part of the work." This engineer is correct. All TRIZ does is generate many innovative (often patentable) ideas. The hard work of determining the correct conditions and materials must still be done. For the diamond cracking example, the 8 atmospheres used for bell peppers would not work. Experiments must be conducted to determine the ideal pressure level and how fast to drop the pressure for the best results. Though the impression that, "All TRIZ has done is generate creative ideas" is correct, the designer has missed the point. For Level 1 problems the use of TRIZ may be irrelevant, but the time required to find ideas for Levels 2, 3 and 4 is significantly greater without TRIZ. Having the best idea means significant time reductions and cost savings during the design process.

Altshuller reasoned that knowledge about inventions should be extracted, compiled and generalized to enable easy access by an inventor in any area. For example, all five inventions mentioned above may be described in the following general way:

> Place a certain amount of pepper, seeds, crystals, etc. into an airtight container, gradually apply increased pressure, then drop the pressure quickly. The underlying principle is that a sudden pressure drop creates an explosion which splits an object. This concept is one of several methods for causing explosions in the TRIZ methodology (see Appendix A).

Consider that the pepper canning method was patented in 1968 and the cedar nut shelling process was not patented until 1986. The patent given for the crystal splitting process was even later. Why should there be these long gaps in development as the discovery of like processes is made independently by different industries?

The general knowledge generated by solutions like these can be organized and used as shown in Figure 10. Inventors should match their problems to similar standard problems. Then possible standard solutions associated with the standard problem can be applied to the specific problem. Via this process, TRIZ accumulates innovative experiences and provides access to the most effective solutions independent of industry. The traditional approach to creativity is to try to jump from "my problem" to "my solution." But there is no repeatable path from "my problem" to "my solution," and any attempt to follow that route could result in never-ending random trials. What seems to be the direct approach, may not get you there in a lifetime.

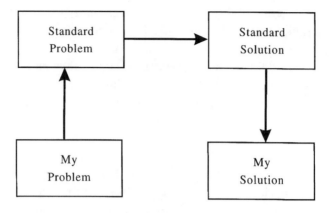

Figure 10. Organizing and using general knowledge.

An alternative approach converts the inventive process to a normal engineering process by taking a given problem to a higher level of abstraction (Figure 11). Stan Kaplan used the example of the quadratic equation in his booklet.[18] The trial and error process is very slow and imprecise. When there is more than one answer that satisfies a requirement, finding an efficient search routine that generates all the satisfying answers is difficult. Once a closed form solution is found, there is no approximation. Your application will probably not use quadratic equations, but this method is analogous to the recommended process — an analogous example to explain analogous thinking.

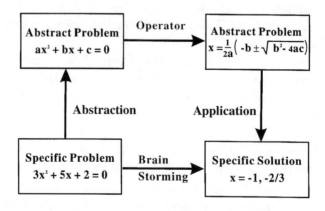

Figure 11. Converting the inventive process to a normal engineering process.

The *Contradiction Table* in Chapter 3 contains 1,201 analogous problems. As a result of using the TRIZ methodology, contradictions become attractive because they define an opportunity for improvement. Tradeoff and compromise will no longer be part of your vocabulary because you will have a way to resolve many contradictions. The ideal design becomes an expectation rather than a dream. Consequently, everything will become a resource, whatever your role in the design process (Table 6).

Table 6. The Ideal Design Expectation

Tradeoff and compromise no longer will be part of your vocabulary.
Everything will become a resource.
Ideality becomes an expectation.
Contradictions are attractive.
Your age will drop in psychological test, (you are off the charts).

And for those looking for the fountain of youth, your "age" will drop on psychological test. Just as children are comfortable with contradictions, you will learn to accept and even appreciate them.

The first step in your transformation begins with the gathering of all the relevant information about your problem. The information will be collected and organized using the Innovation Situation Questionnaire presented in Chapter 2.

> "The greatest task before civilization at present
> is to make machines what they ought to be,
> the slaves instead of the masters of men."
>
> — *Havelock Ellis, 1922*

2 Innovation Situation Questionnaire (ISQ)

System and Problem Information

Professional problem solvers often say that a problem defined well is half solved. The early steps of TRIZ methodology focus on this concept, clearly defining the problem at hand. The innovative team should have a good understanding of the design system surrounding their problem before embarking upon an improvement process. Accordingly, different aspects related to the problem must be systematically documented. The Innovation Situation Questionnaire (ISQ),[19] introduced in this chapter, helps problem solvers with this preliminary task. The ISQ was developed by the Kishinev School of TRIZ in Moldova. Problem solvers are encouraged to record any ideas that occur while completing the questionnaire.

Design or inventive problems are not always clearly defined and all relevant information is not known by the team members. The ISQ makes explicit all the needed information for the individuals working with innovative problems. The ISQ provides the much needed structure for gathering information necessary to reformulate a problem and then break it down into many smaller problems. This crucial procedure of reformulation is discussed in Chapter 3.

Two innovative problems will be used to introduce the ISQ: (1) removing a screw from a bone and (2) improving the speed of a bicycle. The screw is a problem with many constraints. The screw example also illustrates a team that is bouncing back and forth between resolving the problem for one patient today vs. designing a new system to eliminate the problem for future patients. The bicycle problem offers the potential for a radical system change. Each question of the ISQ will be followed by an explanation and the answers for the two cases.

In an actual application, all the questions should be answered in as much detail as possible. A team should spend four to eight hours discussing the content of the questionnaire. This information becomes the database for the different TRIZ tools. It is also very helpful to use generic rather than technology-specific terminology. Thinking with the traditional technical terminology reinforces psychological inertia that TRIZ is attempting to eliminate. Technical terminology often has implied concepts within its vocabulary. To evaporate (a technical term used by a "specialist") is more restrictive than to dry (as used by a 10-year-old child).

Answering these questions stimulates concepts for possible solutions to the innovative problem. Record these solutions as fast as they are created. The combination of this process, and the subsequent formulations of several related problems, yields solutions in 85% of the innovative problems. Concepts related to solution of other tasks and problems should also be recorded. The process of completing the questionnaire gives the problem solver a new set of eyes for viewing the problem or opportunity for improvement.

1. Information about the system you would like to improve/create and its environment.

Use the standard name of your system, if one exists. The explanations in this chapter are based on the screw removal problem and the bicycle speed problem. (The ISQ can also be used for social and nontechnical applications.)

1.1 System name.

Technical system: Screw and Screwdriver (*Screw*)
Technical system: Bicycle (*Bike*)

1.2 System's primary useful function.

A system provides a function when something else is affected. This function can be stated using an active verb describing an object which undergoes some action within the system. Words like *provide* and *produce* are not active verbs. The following function is stated correctly along with constraints: The screw holds the bone surfaces in position during healing. The active verb is *holds*, and the object is *surfaces*. A design constraint for all design concepts is that they not interfere with bone healing (Figure 1).

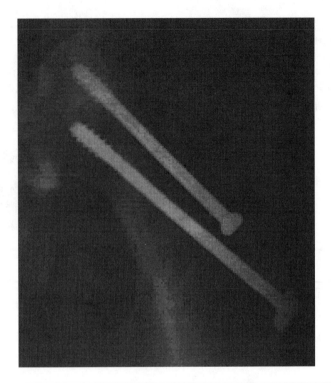

Figure 1. The screws must not interfere with bone healing.

Screw's primary useful function The screw holds bone surfaces in position. The answer to this question will depend upon whether the problem is being developed at the system, super-system or subsystem level. The system is the screw head with a function of transferring torque to the shank. At times, more options are found at the super-system level; this would be the screw, screwdriver, and the bone. Looking at the subsystem, the system would be the hex slot; this is a very limiting problem, since it is a system composed of one element.

Bike's primary useful function The primary useful function of a bicycle is the transportation of a person and small loads over relatively short distances.

1.3 Current or desired system structure.

The structure should be described in its static state (that is, the condition that exists when the system is not operating) and accompanied by a drawing.

Sequentially indicate all subsystems, details and their connections. Super-systems within which the design system operates may also be noted.

Screw The screw has threads (1), a shank (2), a head (3), a metric Allen wrench hole in head (4), and the screw is hollow (5) (Figure 2). The screwdriver (Allen wrench) is also part of the system.

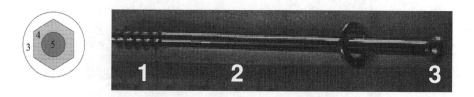

Figure 2. The bone screw.

Bike The bicycle's structure includes the frame (1) and a bearing (2) connected to a shaft (3), on which are fixed foot pedals (4) and a drive sprocket (5). The latter is connected to a chain (6) which is connected to another sprocket (7), which is in turn connected to the shaft (8) and the wheel (9) (Figure 3).

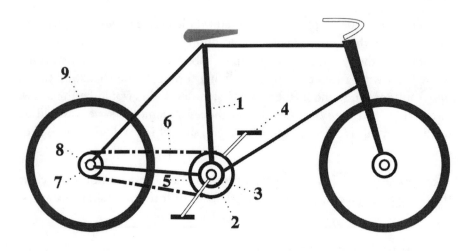

Figure 3. The bicycle.

1.4 Functioning of the system.

Describe how the system functions; that is, how does the system work during the execution of its *primary useful function* and how do its subsystems and elements interact among themselves. Indicate each subsystem sequentially, describing the method and object of the interaction.

Screw The screwdriver is placed in the slot in the head of the screw. For removal, the procedure is to first tighten the screw to break the connection and then to slowly turn it out. (This makes no sense to the patient/author.)

Bike A cyclist rotates foot pedals by pushing down on the down stroke and pulling up (if the foot is strapped to the pedal) on the up stroke. The pedals rotate a sprocket, the sprocket pulls a chain, the chain rotates a second sprocket, the second sprocket rotates a wheel, etc.

"Etc." implies everyone understands the remaining linkages. It is best, however, to describe all the necessary details. After the complete description is written, a frequent response is, "I did not know it worked like that." Many fundamentals of design escape our notice. For example, many people do not realize that spoke tension provides the vertical structure of the bicycle wheel, and not compression as in the spokes of a wooden wagon wheel — an "obvious" detail many could not actually explain.

1.5 System environment.

Many customer-driven organizations investigate the use of environmental conditions that add requirements to the solution of the primary problem. How does the system interact with any super-systems of which it is a part? For this purpose, describing the system's "environment" includes:

- any other system with which the primary system interacts (usefully or harmfully).
- other systems which are located in close proximity to the primary system and may be available for interaction, but do not now directly interact.
- more general systems (super-system, such as public transportation for the bicycle) within which the primary system is a component (subsystem).
- the natural environment (air, water, etc.) surrounding the primary system, such as the human body for the screw.

Screw

- The screwdriver interacts with the screw head and the bone around the head.
- There are many implications and constraints caused by the solution needing to be in a sterile environment and inside a body with little space to maneuver.
- Removing the screw from the leg is part of the super-system of removing foreign materials from the human body.

Bike

- A bicycle interacts with the cyclist, road, air and other vehicles.
- A bicycle is a "neighbor" to other objects in the house or garage where it is kept.
- A bicycle represents a subsystem of a super-system, such as "sporting equipment."
- A bicycle design must be compatible with the "bicycle factory" processes.

2. Available resources.

Problem solvers often overlook the free resources within the environment of application. Innovative designs often take advantage of natural phenomena.

List resources available for use and consider their potential use in eliminating a specified drawback. Use these lists of typically available resources:

- Substance resources
- Field resources
- Functional resources
- Informational resources
- Time resources
- Space resources

See Appendix G for more details on these resources and Chapter 5 for examples.

Screw

- *Functional resources* The body tries to remove foreign bodies.
- *Substance resources* Metal, body bone, tissues, and moisture.
- *Substance resources* Blood and calcium are available in the body.

- *Field resources* A chemical reaction in the body dissolves the screw.
- *Field resources* Mechanical displacement and friction.
- *Field resources* Body temperature.
- *Field resources* Electricity and lights are available in the operating room.

Bike

- *Substance resources* Metal, plastic …
- *Field resources* Free energy can be found in the environment, such as the wind, if it is favorable, and the back draft of a truck.
- *Space resources* Free space in the frame and under the seat. Thread the shift and brake cable through the handlebar.
- *Information resources* The change in sound during the shifting process is an information resource.
- *Functional resources* This would include the possibility of changing gears.

Consider the opportunity of changing existing resources (natural or chemical transformation, processing, accumulation, etc.) for the purpose of eliminating harmful effects. These are called derivative resources.

For example, one resource often considered harmful to the function of washing dishes more efficiently is "grease." By soaking greasy dishes and utensils in sodium bicarbonate, soap is created on the surface where it is needed. This is a derivative resource. There are many derivative resources in our lives that we do not see or we ignore because we view them as drawbacks rather than resources.

3. Information about the problem situation.

3.1 Desired improvement to the system or a drawback you would like to eliminate.

Indicate the cause(s) that led to the problem. (A problem is here defined as anything that is undesirable in the system or the opportunity for improement.) Indicate how this drawback is related to the system's primary useful function and other useful functions. Some typical drawbacks are:

- as required, useful action is absent.
- as required, useful action is implemented ineffectively or incompletely.

- there is a harmful factor in the system such as a harmful action or a harmful result (consequence) of an action.
- required information about an object's condition is absent.
- required information about an object's condition is insufficient.
- the system's degree of complexity is too high.
- the system's cost is too high.
- the magnitude of some characteristic of the system (except the trait of measurement or control) is less than what is required.
- the magnitude of some characteristic of the system (except the trait of complexity or cost) is greater than what is required.
- the value of the system's characteristic of measurement or control is less than what is required.

Screw

- High frictional and cutting forces (for the thread) cause high torque requirements. The primary useful function has already been performed. Subsequent removal is a secondary function, and has failed.

Bike

- A bicycle's speed is limited by the human engine.

3.2 Mechanism which causes the drawback to occur, if it is clear.

If possible, describe the mechanism that causes the drawback, and the conditions and circumstances under which this effect appears. It is always useful to understand the root cause of a drawback. This information is critical to the formulation process explained in the next chapter.

Screw

- Bone has grown around and in the screw. The bone has also bonded to the screw. An Allen wrench hole (hex hole) in the screw head is not the best configuration for high torque. Now there is no longer a hex hole but a near circle. The screw head is below the surface of new bone growth.

Bike

- There are two major causes of low bicycle speed: a) transfer of energy from the cyclist and b) cyclist's aerodynamic resistance. Aerodynamic resistance of the spokes is also a cause, but to a lesser extent.

3.3 History of the development of the problem.

After what events or steps in the system's development did the problem appear? Describe the historic events leading to the drawback and the reasons for it. Consider a path of development that would avoid the problem.

Screw The doctor practices a policy of waiting at least one year before removing a screw. This assures that the break is knitted. After the screw is removed the patient must wait for the hole in the bone to fill before walking without crutches. Though the patient was quite active and much younger than the 70- or 80-year-old patients who usually undergo this operation, the doctor stood by his policy because he lacked data for younger, faster healing patients. Earlier removal could have been easier. (Screws are often just left in the 70- to 80-year-olds.)

Two other options were offered for the initial primary function of having the leg injury heal safely, but their secondary effects were not acceptable. The patient could have:

- prolonged bed rest with leg traction or a full body cast for two months, with possible complications (in 10% of patients).
- no cast because the break did not go all the way through the bone; however, an unusual stress or stumble could complete the break.

It is too late to consider these options, but they and others are available for future patients.

Bike

- The low speed of a bicycle goes back to the bicycle's origination and is related to characteristic features of the design.
- It is possible to change a common bicycle design. For example, a bicycle can be designed where the rider operates from a supine position. This gives rise to new problems in the area of marketing, as well as design. People are accustomed to a conventional design and may not want this significant change. The introduction and resolution of such a secondary problem is undesirable at this point.

3.4 Other problem(s) to be solved.

Is it possible to modify the direction of development so that events leading to a drawback are eliminated? This may cause new drawbacks, but they may be easier to solve.

Screw

- Tools which clamp to the head or core around the screw could work. The clamp requires chipping bone away to clear space for its operation (Figure 4). The coring device will leave a larger hole than the screw would have left if it had been removed successfully during the first attempt (Figure 5). Bone removal weakens the cortex of the bone and requires a longer period of convalescence than when the screw comes directly out. Though these solutions work, a less invasive method is desired and preferably one with a higher likelihood of success.

Figure 4. The bone screw clamp.

Figure 5. The bone screw coring device.

- The screw could be left in place and the patient could be brainwashed to believe that it is OK to have metal in his leg.
- The patient could be given painkilling drugs during cold weather. However, there is evidence that the pain associated with cold weather

does not go away after the screw is removed. This changes the risk/benefit analysis.
- If the original problem were not removing a screw but joining broken bones, then glue could be used as an alternative in the future. To date, glue has not worked independently of some structural support. Biodegradable screws are being developed for some applications.
- If the screw could be used as a conductor for general health improvement by manipulating the magnetic field of the pelvic area, then it would not need to be removed. The screw can be used for monitoring internal bone temperature for a research study, if the patient were willing to participate.

Bike

- An aerodynamic screen can be added to reduce wind resistance. However, this will increase the bicycle's weight and make cycling more difficult. In this case, try solving the primary problem, as well as the secondary one by finding a way to decrease the bicycle's weight while retaining the shields.
- If the problem of lessening aerodynamic resistance cannot be solved, increasing speed by changing the efficiency of the transmission is a useful alternative problem.
- If it is impossible to improve the bicycle, a modification to the function of the bicycle should be considered. A moped is a bicycle with an auxiliary engine. This is not a substitution for human effort, but a supplemental energy source to be used on hills. A small motor scooter used to attain the desired speed is another option.
- A bicycle's function is to move the cyclist and the load. If the function of moving loads is eliminated and the bicycle is used only for sport, increased speed is more easily achieved. Also, if the primary function of the bicycle is exercise, the bike can travel no distance but provide a great workout.
- If it is impossible to increase the bicycle's speed, it becomes necessary to define a new problem. While still traveling at the same speed, the rider could reduce the transit time by using a shorter route. The problem has been changed to route selection. The source of power can be changed for delivery purposes by hiring a trained professional. (Consider the problem of decreasing door-to-door travel time. The airline industry could develop faster passenger planes, but on many

routes the ground transportation, check-in procedure and baggage handling are significant components of total door-to-door time. It is common to work on the traditional issues and not the critical one. The direction in North America continues to be new airports and planes. With the need for expanded security measures, ground time is even longer. Alternatively, some countries have replaced the air routes between major cities with rapid transit trains, because the door-to-door time is less. It is very important to work on the bottleneck that detracts from performance.)
- Public transportation is a super-system of the bicycle. Some cities design buses to allow for transport of cyclist and bicycle. Efficient transportation by bicycle can probably be used in other super-systems, such as community health and nonpublic transportation (police patrol).

For any particular case, not every alternative will yield viable problems with potential solutions. Still, it is worthwhile to examine all opportunities. Even if they do not lead immediately to a new solution, it is possible that these alternative problems will prove useful at later stages of the solution process. New products may be conceived at this stage. Often, one project becomes two — one for today and one for the future.

4. Changing the system.

4.1 Allowable changes to the system.

Evaluate and describe the degree of possible change to the system that is achievable as a result of the problem solving process. Basically, the degree of possible changes is dependent upon:

- the current stage of the system's production (in development, working samples, pilot production, mass production, existing and well-established technology, etc.).
- losses (direct and indirect) caused by this negative effect.
- possible profit and other benefits from solving the problem.

Which of the following statements best describes the innovative problem?

Complete changes are possible, including creating a new product and/or technology.
Screw
Any tool for removal is possible.

Major changes are possible within limits defined by cost, development, equipment and compatibility with marketing strategies (same customers/market size).
Bike
There is concern about future sales following an image change.

Only small changes are possible; options are restricted by the necessity to retain the existing technology, existing obligations, customer requirements, etc. Be specific when defining restrictions.
Bike
Major changes to the manufacturing process would be expensive.

It is possible to make only minimal changes. Indicate exactly why.
Screw
Concern about the patient's well-being.

Indicate which technical, economic or other characteristics may be changed. What are the constraints to these changes?

Screw

- A new tool for removing the screw may be designed, but it must withstand high temperatures for sterilization and fit space requirements to minimize invasiveness of the procedure.

Bike

- Complete changes to the design and manufacturing technology are possible. The only limitation is that it still look like a bicycle so as not to deter customers.

4.2 Limitations to changing the system.

Indicate what can or cannot be changed in the system. Which technical, economic or other characteristics should:

- remain constant?
- not decrease?
- not increase?

Explain the reasons for the imposed restrictions. If possible, indicate conditions under which these restrictions can be removed. If removal of the restrictions causes new (secondary) problems, evaluate whether it is better to try to solve these problems.

Screw

- The environmental conditions cannot be changed, that is, it is sterile inside the body. A sterile environment is necessary for the safety of the patient. The safety restrictions cannot be ignored for the patient or the medical staff.

Bike

- The bicycle cannot be replaced by another means of transportation, but the mode of transportation can be influenced by environmental conditions. Some other means of transportation may be necessary in the winter because of cold temperatures or deep snow. Safety and convenience for the cyclist must not be reduced. The reputations of organizations building and selling bicycles are based on the convenience and safety of past designs. Safety does not, however, seem to be an issue for people who race bicycles. The amount of inconvenience the rider will tolerate also depends upon the design intent of the bicycle. For the competition bicycle, safety restrictions can be partially ignored.

5. Criteria for selecting solution concepts.

5.1 Desired technological characteristics.
5.2 Desired economic characteristics.
5.3 Desired timetable.
5.4 Expected degree of novelty.
5.5 Other criteria.

Indicate those parts of the system that need to be changed to achieve the desired characteristics delineated above. What changes need to be made (both qualitative and quantitative) to achieve these characteristics? How and why will these changes affect the drawback? Indicate the basis of the criteria by which possible solutions will be evaluated.

Screw

- Reduce frictional forces, increase torque potential to the head, and reduce the effort of cutting new thread in bone. The effort required to turn the screw is decreased by reducing the friction and easing the cutting process. A higher torque could be applied without slipping or causing damage to the head.
- Is there an organic version of machine cutting oil? Can it resemble penetrating oil which would move down along the screw to reduce friction without damage to the bone? These ideas are recorded here as a note because the response simulated the concept. In this way, it is connected to the original stimulus. However, the screw has usually bonded with the bone and would require a different effect.
- Other criteria for a new design should consider:
  - the time required to develop a working model
  - the time required to remove the screw
  - the length of the incision required to use the tool
  - the likelihood that the screw will be removed.

Bike

- It is necessary to reduce aerodynamic resistance of the bicycle and the cyclist. The reduction in aerodynamic resistance will increase the speed with no change or a decrease in effort. Design criteria should include:
  - the percent that speed is increased
  - the size of the investment needed to implement the change
  - the likelihood of receiving a patent
  - the closeness to classic bicycle appearance
  - the likelihood of increased sales.

6. History of attempted solutions to the problem.

6.1 Previous attempts to solve the problem.

In documenting the history of previous attempts to solve the problem, try to define the reason that earlier attempts have failed.

6.2 Other system(s) in which a similar problem exists.

Name systems in which an analogous problem exists, and then ask:
- Has the problem been solved elsewhere?
- Is it possible to apply such a solution to the current problem?
- If it is impossible, why? What are the limitations?

Screw

Analogous problems include:

    a. bolts rusted in metal.
    b. wood screws with stripped heads.
    c. nails driven in living tree years earlier.

Solutions:

    a. Use an easy-out for bolts. Drill a hole in the exact center of a right-threaded bolt (the normal direction). By turning a tapered left-threaded screw until tight and then continuing to turn it to the left, the right-threaded bolt will back out.
    b. Use a hack saw to cut a new slot in the head of the screw.
    c. New tree growth around the head of the nail can be removed for better purchase area.

Limitation:

All these methods could work for the screw in the bone, but because the screw removal must occur in a sterile environment as quickly as possible and without leaving any foreign materials in the leg or bone, drilling and cutting should be minimized. The head-shank connection is the weakest link in the distribution of torque along the length of the screw. Removal will be very difficult if the head breaks off.

Bike

Analogous problems include:

    a. increasing the speed of roller skates.
    b. increase the speed of a row boat (where there is wind and water resistance).

Solutions:

    a. Change the system to roller blades with new materials and bearings.

b. Change the shape of the hull and the oars. Change the pattern for rowing.

Limitation:

The new system must look like the classic bicycle.

Many TRIZ practitioners find that The Innovation Situation Questionnaire provides all the guidance they needed to solve a particular problem. Others continue with the next step in the TRIZ methodology. Now that you have recorded all the information relating to the innovative problem, you will be introduced to the Problem Formulator which restructures the primary problem into many smaller problems (Chapter 3). For large projects, several different formulations may be needed. Looking at the problem in such detail may seem costly, but consider the waste and expense that characterize poorly planned design efforts. Everyone agrees good planning pays off, but resources are rarely made available for planning.

> *"He that will not apply new remedies must expect new evils for Time is the greatest innovator."*
>
> *— Francis Bacon*
> *from "of innovations," Essays, 1625*

3 Problem Formulation

Problem Formulation Process

This chapter presents the *formulation process* for constructing a simple cause and effect graph to show the linkage between the primary drawback in a system and the primary useful function.[20-22] TRIZ methodology uses the terms *harmful function* and *useful function* to indicate the drawback and primary function, respectively. Note that, in this context, "function" has a very loose definition and should not be confused with the definition used for Chapter 2's ISQ analysis. This more general definition describes function as "anything you want it to be." Many "functions" are events, such as an explosion or sudden metamorphosis of a system part. The formulation process accommodates functions that belong in this category.

To construct the graph, begin with either the Primary Harmful Function (PHF) or the Primary Useful Function (PUF). If you start with the PHF, try to identify the functions that can be linked to the PUF. Linkage is considered complete when there is at least one path from PHF to PUF.

The graph provides a useful image of the many interrelated smaller problems included within the innovative problem. To effectively resolve the innovative problem (eliminate PHF), formulate the cause and effect relationships of all the related problems. Once all of these relationships are identified, you may discover that resolving only one of these secondary formulated problems will resolve the primary problem. Begin with the problem that has the most significant impact on the overall system.

The previous chapter identifies information relevant to the resolution of the primary problem. The problem formulation builds upon this information.

The next step in problem formulation involves wording the problems in precise phrases that describe the interaction between useful and harmful system functions. These specially formatted phrases must include at least two functions.

Getting Started

Eight questions help identify links between functions. Asking each of the questions assures that no relationships are forgotten. There are three links between harmful (**HF**) and useful functions (**UF**):

1. **UFn** causes ━━▶ **HFn**
2. **UFn** is introduced to eliminate ─┼─▶ **HFn**
3. **UFn** is required for ━━━▶ **UFn+1**

These three links lead to eight questions, four relating to useful functions and four to harmful functions.

Four Questions for Useful Functions

1. Is this useful function *required for* another useful function(s)?
   UFn is required for ━━━▶ **UFn+1**
2. Does this useful function *cause* any harmful effect(s)?
   UFn causes ━━▶ **HFn**
3. Has this useful function been introduced to *eliminate* a harmful effect(s)?
   UFn is introduced to eliminate ─┼─▶ **HFn**
4. Does this useful function *require* another useful function(s) in order to perform?
   UFn-1 is required for ━━━▶ **UFn**

Four Questions for Harmful Functions

5. Does this harmful function *cause* another harmful function(s)?
   HFn causes ━━▶ **HFn+1**
6. Is this harmful function *caused by* another harmful function(s)?
   HFn-1 causes ━━▶ **HFn**
7. Is this harmful function *caused by* a useful function(s)?
   UFn causes ━━▶ **HFn**
8. Has a useful function been introduced to *eliminate* this harmful function(s)?
   UFn is introduced to eliminate ─┼─▶ **HFn**

Questions 1 through 4 relate to any useful function. The first three questions identify events that follow from the useful function. Only Question 4 relates to the PUF, since the PUF represents the end of the process. Once the first cycle (which explores the function most immediately related to the PUF) is complete, all four questions become relevant.

An affirmative answer to any question identifies another function or another link in the model. These relationships are depicted with a flow chart.

Some of the information was previously identified in ISQ answers. More than one question can have an affirmative answer. Each affirmative answer can result in more than one link. It is important to explore the appropriate set of four questions for every function in forming the link model.

The PUF is identified in the second inquiry of the ISQ (Chapter 2, 1.2), "System's primary useful function." The source for the PHF is the third ISQ question (Chapter 2, 3), "Information about the problem situation." You may begin the process with either the PUF or the PHF. Use each affirmative answer to lead you to another function, either useful or harmful. A flow chart of the questions clearly depicts the inquiries that this thorough analysis requires (Figures 1 and 2). A question for one function type generates a set of four questions for each newly considered function.

Two examples, using PUF and PHF respectively, will be used to show the formulation process. A furnace application will demonstrate problem formulation starting from a PUF. Then the bone screw will be used again to demonstrate an application beginning with a PHF. In order to visually identify harmful functions (HF) and useful functions (UF) easily when formulating the problem, drawing flow charts and writing problem statements, useful functions will be in parentheses (UF) and harmful functions will be underlined and in brackets [HF]. Different shapes for useful functions (white circles) and harmful functions (gray squares) will be used to facilitate reading of the flow chart. Different lines will be used for each of the three links: a black line indicates one function is required for the other function, a gray line indicates one function causes another function, and a slashed black line indicates one function eliminates another function.

Furnace Problem

An organization designs furnaces for processing ores. The customer has requested a furnace that is fast, efficient and compact. The ability to withstand high temperatures is a critical design parameter. Past experiences with metal extracting furnaces would be described as:

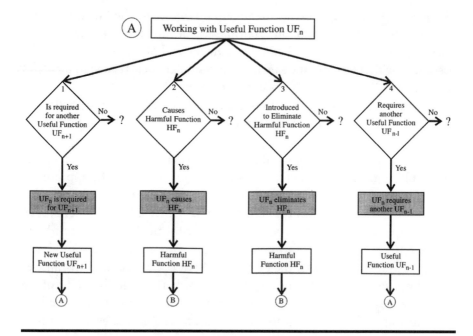

Figure 1. Working with a useful function, UF_n

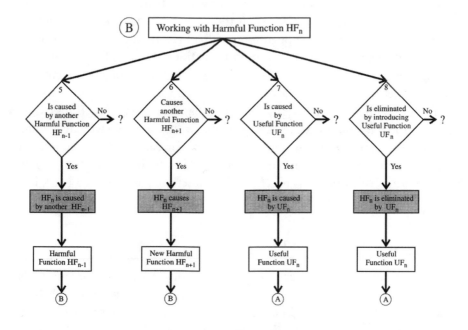

Figure 2. Working with a harmful function, HF_n

Metal is extracted from ore by melting it in a high-temperature furnace. To cool the brick walls of the furnace, water is pumped through pipes enclosed within the walls. If a pipe cracks, water may leak through the brick wall into the furnace, resulting in an explosion.

The PUF is extracting metal. What function does extracting metal require? To keep this example simple for the purpose of illustration, only (melting ore) will be used as the "yes" answer to Question 4:

(Melting ore) is required for (extracting metal).

The relationship is shown as a flow chart in Figure 3.

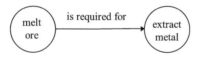

Figure 3. Flow chart for melting ore/extracting metal.

It is important to emphasize that (melting ore) only has one meaning. It does not include any information or assumptions about containers, temperature, atmosphere or other conditions. These other conditions may be additional functions, and they should be diagrammed and explored accordingly. Creating a flow chart of the system before applying the Formulator clarifies these separate functions. For (melting ore), only Question 4 brings you to a new function:

(High temperature) is required for (melting ore).

This new function is added to the left-hand portion of the flow chart in Figure 4. An exhaustive set of problem statements requires a detailed list of functions.

Including melting ore adds three more problem statements.

Problem formulation articulates all the links between the two functions in order to identify all existing relationships. (High temperature) has a "yes" response to Question 2:

(High temperature) causes [<u>overheating furnace walls</u>].

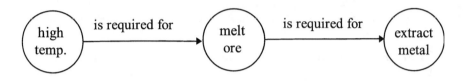

Figure 4. High temperature added to flow chart shown in Figure 3.

Since [overheating furnace walls] is a harmful function, we now begin to ask Questions 5 through 8. A "yes" to Question 8 yields:

(Cooling walls) eliminates [overheating furnace walls].

We can also form statements in response to Questions 5, 6 and 7. For the sake of brevity, the example will only follow one possible chain:

(Moving water through pipes) is required for (cooling walls).

(Pumping water through pipes) is required for (moving water through pipes).

(High pressure) is required for (pumping water through pipes).

(High pressure) causes [leakage of water into furnace].

[Cracks in pipe] causes [leakage of water into furnace].

This process can now be continued for [cracks] and [leakage of water into furnace]. For the purpose of illustration, only [leakage of water into furnace] will be explored:

[Leakage of water into furnace] causes [explosion].

Several of the questions could have had "yes" answers, but were not considered in order to simplify the model for learning purposes.

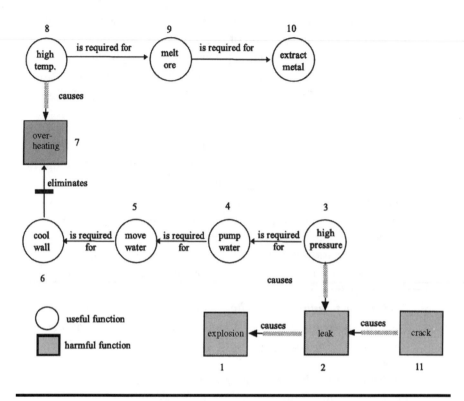

Figure 5. The numbers at each node refer to the problem statements below.

Developing Problem Statements

Accepting [explosion] as the problem may lead you to overlook the other 21 subproblems that were formulated. In Figure 5, the numbers at each node refer to the problem statements listed below. The links in the diagram suggest two types of problem statements: the ***preventive*** statements for harmful functions and the ***alternative*** statements for useful functions. An additional problem statement for each node can be expressed as either a benefit from a harmful function, an enhancement to a useful function, or the resolution of a contradiction.

1a. ***Find a way to eliminate, reduce or prevent*** [explosion] under the condition of [leakage of water into furnace].
1b. ***Find a way to benefit from*** [explosion].

2a. ***Find a way to eliminate, reduce or prevent*** [leakage of water into furnace] under the condition of (high pressure) and [cracks in pipe].
2b. ***Find a way to benefit from*** [leakage of water into furnace].
3a. ***Find an alternative way of*** (high pressure) that provides (pumping water through pipe) and does not cause [leakage of water into furnace].
3b. ***Find a way to enhance*** (high pressure).
3c. ***Find a way to resolve contradiction:***
   (High pressure) should provide (pumping water), and should not cause [leakage of water into furnace].
4a. ***Find an alternative way of*** (pumping water through pipe) that provides (moving water through pipe) and does not require (high pressure).
4b. ***Find a way to enhance*** (pumping water through pipe).
5a. ***Find an alternative way of*** (moving water through pipe) that provides (cooling furnace walls) and does not require (pumping water through pipe).
5b. ***Find a way to enhance*** (moving water through the pipe).
6a. ***Find an alternative way of*** (cooling furnace walls) that prevents [overheating furnace walls] and does not require (moving water through pipe).
6b. ***Find a way to enhance*** (cooling furnace walls).
7a. ***Find a way to eliminate, reduce or prevent*** [overheating furnace walls] under the condition of (high temperature) that does not require (cooling furnace walls).
7b. ***Find a way to benefit from*** [overheating].
8a. ***Find an alternative way of*** (high temperature) that provides (melting ore) and does not cause [overheating furnace walls].
8b. ***Find a way to enhance*** (high temperatures).
8c. ***Find a way to resolve contradiction:***
   (High temperature) should provide (melting ore), and should not cause [overheating].
9a. ***Find an alternative way of*** (melting ore) that provides (extracting metal) and does not require (high temperature).
9b. ***Find a way to enhance*** (melting ore).
10a. ***Find an alternative way of*** (extracting metal) that does not require (melting ore).
10b. ***Find a way to enhance*** (extracting metal).
11a. ***Find a way to eliminate, reduce or prevent*** [cracks in pipe].
11b. ***Find a way to benefit from*** [cracks in pipe].

From a system's point of view, there are three levels of problem definition. Problem Statements 1a, 1b, 2a, 2b, 3a and 3b are at the subsystem level and require specific, detailed solutions. Problem Statements 8a, 8b, 9a, 9b, 10a, 10b, 11a and 11b require major changes to the system, which may include system replacement and represent a long term solution. Problem statements that lie between these two extremes (Statements 4a, 4b, 5a, 5b, 6a, 6b, 7a and 7b) involve significant changes of a less radical nature — these may be useful in solving the current crisis. The formulation process offers a complete list of problem statements. This thorough analysis gives the problem-solving team a variety of directions to follow for developing the best solution.

Flow charts, such as Figure 5, can be used to find visual clues for selecting profitable problem statements.

A well-formulated problem is a problem that is nearly solved. Often when developing problem statements, solutions to a problem become obvious because the relationships between functions are more easily observed than in a traditional single-problem structure (for example, [cracks in pipes] cause [explosions]). Professional knowledge and common sense are often sufficient to resolve a problem statement. Consider problem Statement 2a,

> **Find a way to prevent** [leakage of water into furnace] under the condition of (high pressure) and [cracks in pipe].

Why does the water leak when there is high pressure and a crack in the pipe? Because the pressure outside the pipe is less than inside the pipe. Therefore, if the pressure inside the pipe were less than the pressure outside, there would be no leak. Is this possible? Yes, by using a vacuum pump to move the water. Question 4a, "**Find an alternative way of** (pumping water through pipe) that provides (moving water through pipe) and does not require (high pressure)," could have also triggered this concept. This questioning and reasoning process shows how a well-formulated problem easily leads to practical, inexpensive solutions. Continue to pursue solutions to the other problems, but remember that highly technical solutions are not always required.

This solution to [leakage of water into furnace] breaks the link between (high pressure) and (pumping water through pipes). In this way, the four nodes furthest from the PUF are eliminated. The closer the problem statement is to the PUF, the more major the required changes to the system.

The problem formulation process offers directions for finding solutions that can be implemented today, tomorrow and in the more distant future.

The problem statements developed for the furnace point in several evolutionary directions for the process of extracting metal from ore.

Find an alternative way of (extracting metal) that does not require (melting ore). This problem statement suggests looking for nonheating processes. A chemical process may be possible.

The formulation process can be used for any problem, not just for technical issues. To adapt this tool for a service application, simply replace the word "function" with the word "task."

For systems that have many failure modes, use Quality Function Deployment (QFD) and/or Analytic Hierarchy Process (AHP) to prioritize failures. This prioritization identifies the most important HF (or PHF), which will become your starting point for the formulation process. Large problems may generate hundreds of problem statements. These should be grouped so that allocation of resources to the most promising issues is facilitated.

If your organization lacks the resources needed to look at all problem formulations, use the following guidelines to shorten the problem selection process:

1. Select the problem with the best cost/benefit ratio.
2. The more radical the problem, the greater the potential benefit.
3. It is better to eliminate a harmful cause, than to mitigate results.
4. The level of difficulty involved in implementation of a solution should be a factor in problem selection. Too radical a solution may prove unacceptable, depending on an organization's culture and psychological inertia.

Not all problems are as simple as the furnace example. If the eight questions include the following hypothetical answers, the problem statements are more involved than those in our example.

(PUF) requires (UF1).
(PUF) requires (UF2).
(UF2) causes [HF1].
(UF3) causes [HF1].
(UF4) eliminates [HF1].
(UF4) causes [PHF].

As in the earlier example, the numbers by the nodes in Figure 6 indicate the associated problem statement.

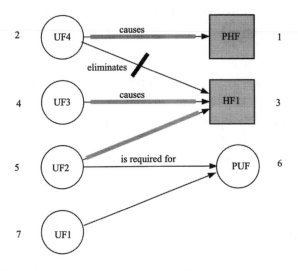

Figure 6. Generic network of relationships.

1a. ***Find a way to eliminate, reduce or prevent*** [PHF] under the condition of (UF4).
1b. ***Find a way to benefit from*** [PHF].
2a. ***Find an alternative way to provide*** (UF4) that eliminates, reduces or prevents [HF1] but does not cause [PHF].
2b. ***Find a way to enhance*** (UF4).
2c. ***Find a way to resolve contradiction:***
   (UF4) eliminates [HF1] and should not cause [PHF].
3a. ***Find a way to eliminate, reduce or prevent*** [HF1] under the condition of (UF3) and (UF2) and does not require (UF4).
3b. ***Find a way to benefit from*** [HF1].
4a. ***Find a alternative way to provide*** (UF3).
4b. ***Find a way to enhance*** (UF3).
5a. ***Find an alternative way to provide*** (UF2) that provides or enhances (PUF) and does cause [HF1].
5b. ***Find a way to enhance*** (UF2).
5c. ***Find a way to resolve contradiction:***
   (UF2) should provide (PUF) and should not cause [HF1].
6a. ***Find an alternative way to provide*** (PUF) that does not require (UF1) and (UF2).
6b. **Find a way to enhance** (PUF).

7a. ***Find an alternative way to provide*** (UF1) that provides or enhances (PUF).
7b. ***Find a way to enhance*** (UF1).

These new patterns in the flow chart represent two different cases. The first pattern shows a function that is necessary to provide another useful function, but also causes a harmful function. The second pattern shows a useful function eliminating a harmful function, but creating a new harmful function.

Formulating the Screw Removal Problem

Let's look at problem formulation again, using the example of the screw in the patient/author's leg. This time we will begin with the PHF. Additional information that would normally be part of the ISQ is:

> A crack formed in the neck joining the leg femur to the ball. The crack was through the narrowest diameter of the neck. It was caused by impact directed upward toward the ball from the lower right corner of Figure 7.

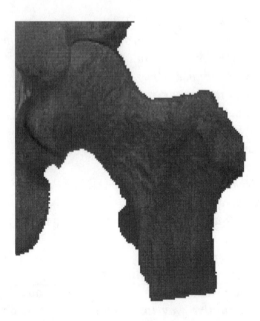

Figure 7. A model of a left femoral neck.

The screws in Figure 8 are 96 and 73 mm long. The screw with the washer was removed after one year. The head of the other screw was stripped during attempted removal during the same operation, and it was left in place. The author/patient decided to apply TRIZ to generate more options than the doctor could offer. Leaving the screw in is not an acceptable solution to the author/patient.

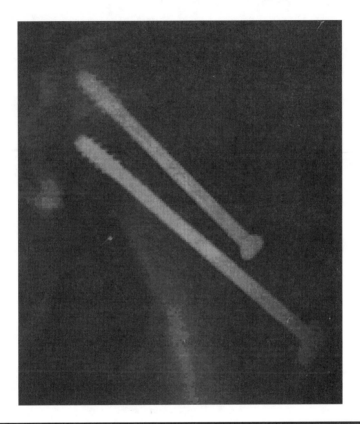

Figure 8. Stainless steel screws in femoral neck.

The stainless steel screw is 73 mm long and has a 16 mm thread. The thread is self-cutting in either direction. The shank is 4.5 mm in diameter (Figure 9). (Shown is the 96 mm screw.)

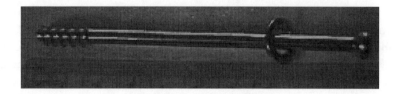

Figure 9.   96 mm stainless steel screw with self-cutting thread.

The threads of the screw are designed to cut the bone going in or out (Figure10). Note the 2 mm hole, used by a wire inserted in the bone as a guide for placing the screw.

Figure 10.   The screw threads are designed to cut the bone during insertion or removal.

The diameter of the head is 7 mm with a hexagonal hole 4.5 mm across the face (Figure 11). Again, note the 2 mm hole. This hole may be a resource.

The desired improvement to the system is removal of the remaining screw from author/patient's left femoral neck. The PHF is [screw cannot turn out]. The problem formulation process identifies the following links:

[Screwdriver head slips in slot] causes [screw cannot turn out].

[Slot is rounded] causes [screwdriver head slips in slot].

Figure 11. The 2 mm hole may be a resource in removing the screw.

(Large frictional force) causes [slot is rounded].

(Large frictional force) is required for (screw integrated into bone).

(Large frictional force) causes [screwdriver slips in slot].

(Bone growth around threads and shaft) requires (large frictional force).

(Heal crack in bone and fill hole) requires (bone growth around threads and shaft).

The structuring of this formulation is simple. Ideas were recorded as they were conceived while reading the problem statements.

A flow chart depicting the function relationships (Figure 12) yields the following problem statements:

1a. ***Find a way to enhance*** (heal crack in bone and fill hole).
1b. ***Find an alternative way of*** (heal crack in bone and fill hole) that does not require (bone growth around threads and shaft).
2a. ***Find a way to enhance*** (bone growth around threads and shaft).
2b. ***Find an alternative way of*** (bone growth around threads and shaft) that provides (heal crack in bone and fill hole) and does not cause (large frictional force).
3a. ***Find a way to enhance*** (large frictional force).
3b. ***Find an alternative way to provide*** (large frictional force), which provides or enhances (screw integrated into bone), which does not cause [screwdriver slips in slot] and does not require (bone growth around threads and shaft).
3c. ***Find a way to resolve contradiction:***

(Large frictional force) should provide (screw integrated into bone) and should not cause [screwdriver slips in slot].

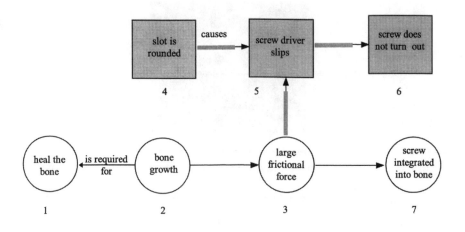

Figure 12. The flow chart depicts the preceding problem statements.

Concepts: lubricate, vibrate, reduce dimension of screw, soften bone, dissolve screw, dissolve threads. At this point, the patient became quite concerned as he pictured the application of each of these concepts — to his leg! Although someone working on a problem solution is not usually so emotionally involved, this situation provided a dramatic example of the Voice of the Customer saying, *"no way."*

4a. ***Find a way to benefit from*** [slot is rounded].
4b. ***Find a way to prevent*** [slot is rounded] under the condition of (large frictional force).
5a. ***Find a way to benefit from*** [screwdriver slips in slot].
5b. ***Find a way to prevent*** [screwdriver slips in slot] under the condition of [slot is rounded] and (large frictional forces).

Concepts: use left-handed easyout, vice grip, glue, cut new slot, apply torque to tip of screw.

6a. ***Find a way to benefit from*** [screw cannot turn out].
6b. ***Find a way to prevent*** [screw cannot turn out] under the condition of [screwdriver slips in slot].

7a. **Find an alternative way to provide** (screw integrated into bone), which does not require (large frictional force).
7b. **Find a way to enhance** (screw integrated into bone).

Problem formulation, together with the ISQ, is a powerful innovation method. It has produced a number of industrial breakthroughs, along with several patents.

Rockwell International wanted to improve the design of golf cart brakes. It identified 21 linking statements that resulted in 42 problem statements. The net result was an effective braking system containing significantly fewer parts.

Allied Signal used TRIZ to redesign a containment ring for compressors on jet planes. There were nine link statements that led to innovative designs. Parts of this case appears in Appendix F.

The ISQ and Problem Formulator material are not considered part of "classical" TRIZ, but were created to eliminate limitations in the classical methodology. These two elements are now important components of the TRIZ toolbox.

The next chapter presents a third feature of the innovator's toolbox. This component was one of the first tools developed by Altshuller: the resolution of conflicts using the 40 Principles or the separation principles.

> "Almost all men are intelligent. It is the method that they lack."
>
> — F.W. Nichol

4 Contradiction

When to Use Contradiction Analysis

Contradiction analysis is a powerful method of looking at your problem with new eyes. Once you have gained this fresh perspective, the Contradiction Table becomes your tool for generating numerous solution concepts. If your problem fits into the parameters outlined below, you may be well on your way to finding a variety of solutions that are both creative and effective.

The Origin of Contradiction Analysis

Often a challenging problem can be expressed as either a technical or physical contradiction. The use of the words *technical* and *physical* is somewhat arbitrary, but these are the accepted terms in TRIZ literature.

A challenge to be overcome is called a technical contradiction when known alternatives available to improve one aspect of a design do so at the expense of another aspect of the design. In other words, a technical contradiction exists if improving parameter "A" of the system causes parameter "B" to deteriorate (as a container becomes stronger it becomes heavier). Some technical contradictions are: (1) faster ice skates (racing skates) are less maneuverable (figure skates), and (2) faster automobile acceleration reduces fuel efficiency.

A physical contradiction exists if some aspect of a product or service must have two opposing states (the product is hot, and it is cold).

Physical Contradictions: A Closer Look

A car owner wants a small car for city driving and parking. On the other hand, he or she wants the car to be as large as possible for easy egress and comfort. In 1995, Honda challenged their engineers to forget traditional body styles and to design a car around these comfort and egress requirements. The difficulty was that the car also needed to be parked easily in Tokyo. The design team developed a car approaching the shape of a ball. This design could be both small on the outside and large on the inside. For this physical contradiction, the equivalent technical contradiction is that to make the car large also makes it more difficult to park.

If the problem matches either the physical or technical contradiction format, then the material in this chapter *may* be useful. (**Note:** The word *may* is used because there is another condition which is explained in the technical contradiction section.)

Structuring an Innovative Problem into a Contradiction

Understanding how to structure your problem as a contradiction is an essential step in this analysis. The information contained in the ISQ of Chapter 2 organized your knowledge about your problem situation in a format that helps to systematically explore the *solution space*. This questionnaire includes a series of questions, such as "What is the primary useful function of the system?" and "What do you want to improve?" Some of the answers to these questions could already contain contradictions.

The following two contradiction examples come from the cases discussed in Chapter 2. "Remove the screw from the bone" and "Increase the bicycle speed" were the answers to Question 1 in the ISQ ("What do you want to improve?").

Technical contradictions are usually caused by the current design consideration or physics. But be careful, the constraint of physics may not be real. As explained in Chapter 2, many people believe centrifugal force always throws material away from the axis of rotation. The little-known Weissenberg effect, however, describes materials that grow along the axis of rotation. Today's constraints may only be the result of lack of knowledge. Looking at the physical effects in Appendix A widens the available knowledge base and creates more options.

Question 3 in the ISQ ("What is the cause of the problem?") provides the technical contradiction for the bicycle: increasing speed requires more physical power.

Question 4 articulates another technical contradiction for the bicycle: reduced wind resistance causes more weight. (This situation also contains a physical contradiction: having the windshield for reduced wind resistance and not having the windshield for reduced physical power).

The problem formulation process of Chapter 3 develops all possible problem statements related to your challenge. This powerful process can also be used for nontechnical problems. A technical contradiction is formed when a desired function A requires a function C which damages B or causes an undesired function B. From Chapter 3, we could have *function A requires function C and function C causes function B*. The following statements are from this technical contradiction:

Find an alternative way of (C) which provides or enhances (A) and does not cause [B].

Find a way to eliminate, reduce or prevent [B], under the condition of (C).

Find a way to enhance (C).

Find a way to resolve contradiction:
(C) should provide (A), and should not cause [B].

Find an alternative way to provide (A), which does not require (C).

Find a way to enhance (A).

Find a way to eliminate, reduce or prevent [B], under the condition of (C).

Find a way to benefit from [B].

For the screw problem, we start with, "(Large frictional force) is required for (screw integrated into bone) and causes [screwdriver slips in slot]." Reformulated, this is, "***Find a way to resolve CONTRADICTION:*** (Large frictional forces) should provide (screw integrated into bone), and should not cause [screwdriver slips in slot]." The improvement of screw integrating into the bone makes it more difficult to remove the screw. The Problem Formulator process reveals the technical contradictions contained in a particular challenge.

The Contradiction Table: Improving the Normal Problem Solving Process

Traditional problem solving builds on past experiences.

> Humans solve problems by analogic thinking. We try to relate the problem confronting us to some familiar, standard class of problems (analogs) for which solutions exist. If we draw upon the right analogy, we arrive at a useful solution. Our knowledge of analogous problems is the result of educational, professional, and life experiences.[23]

Table 1. 39 Parameters

1. Weight of moving object	21. Power
2. Weight of non-moving object	22. Waste of energy
3. Length of moving object	23. Waste of substance
4. Length of non-moving object	24. Loss of information
5. Area of moving object	25. Waste of time
6. Area of non-moving object	26. Amount of substance
7. Volume of moving object	27. Reliability
8. Volume of non-moving object	28. Accuracy of measurement
9. Speed	29. Accuracy of manufacturing
10. Force	30. Harmful factors acting on object
11. Tension, pressure	31. Harmful side effects
12. Shape	32. Manufacturability
13. Stability of object	33. Convenience of use
14. Strength	34. Repairability
15. Durability of moving object	35. Adaptability
16. Durability of non-moving object	36. Complexity of device
17. Temperature	37. Complexity of control
18. Brightness	38. Level of automation
19. Energy spent by moving object	39. Productivity
20. Energy spent by non-moving objects	

What if we have never encountered a problem analogous to the one we face? This obvious question reveals the shortcomings of our standard approach to inventive problems. A table of conflicts (Contradiction Table, Appendix D) between 39 design parameters (Table 1) answers this question of how we can face an unfamiliar conflict by offering 1201 generic problems that were solved using at least one of 40 generic principles (Appendix C and Table 2).

Many problem solvers try going directly from problem to solution through trial and error. Looking at an analogous standard problem and its associated standard solution is a more efficient approach. Through the Contradiction Table, TRIZ methodology opens up the world patent base for identifying principles that may offer possible solutions. Structuring the design problem as a contradiction allows the problem solver to fit a problem into the structure of the TRIZ Contradiction Table. The table offers several principles frequently used to solve analogous problems. The problem solver can now concentrate on adapting these likely standard principles to the innovative problem.

Table 2. 40 Principles

1. Segmentation
2. Extraction
3. Local quality
4. Asymmetry
5. Combining
6. Universality
7. Nesting
8. Counterweight
9. Prior counter-action
10. Prior action
11. Cushion in advance
12. Equipotentiality
13. Inversion
14. Spheroidality
15. Dynamicity
16. Partial or overdone action
17. Moving to a new dimension
18. Mechanical vibration
19. Periodic action
20. Continuity of useful action
21. Rushing through
22. Convert harm into benefit
23. Feedback
24. Mediator
25. Self-service
26. Copying
27. An inexpensive short-lived object instead of an expensive durable one
28. Replacement of a mechanical system
29. Use a pneumatic or hydraulic construction
30. Flexible film or thin membranes
31. Use of porous material
32. Changing the color
33. Homogeneity
34. Rejecting and regenerating parts
35. Transformation of physical and chemical states of an object
36. Phase transition
37. Thermal expansion
38. Use strong oxidizers
39. Inert environment
40. Composite materials

There are contradictions in all that we see and every thought we have, but we usually do not explore them. (For example, the entrance to a bus should be large for easy egress and small for maximum seating.) If we did, we might ask, "What if?" This "What if" is not the same as asking, "What is wrong?" It is a means of discovering a new perception of known phenomena that we often take for granted. Related questions such as, "What is affected by the current 'design' decision?" or "What if we used another system?" help us uncover solutions that we habitually overlook. Before exploring how TRIZ facilitates fresh thinking about your problem, let's take another look at physical and technical contradictions.

Technical Contradictions

An inventive problem contains at least one contradiction, according to Altshuller. This definition was based on his observations of numerous patents. As a result of Altshuller's work, each scientist does not have to look at all patents from all disciplines. A problem need only be put in the format of the contradiction table. Which of the 40 Principles we use is determined by which parameter is being improved and which is being degraded. If the order of the parameters is reversed, different principles are recommended. For some problems, the contradiction is valid in either direction because both parameters need improvements. In this case, use the table twice.

Representing your contradiction as a combination of two of the 39 Parameters requires a broad interpretation of the parameters. The screw problem has "screw integrated into the bone" improving or mending the bone. There is nothing in the 39 Parameters in Table 1 (or in the definitions of Appendix B) that comes close to this phrase. Stretching the imagination, however, "screw integrated into the bone" could be described as improving the "volume of a nonmoving object." This is Parameter 8. Using the same type of transformation in thinking, the "screwdriver slips in slot" is a degradation of "force," which is Parameter 10.

Table 3.   Principle 7 — Nesting

a. Contain an object inside another, which in turn is placed inside a third object.
   Example: Mechanical pencil with lead storage

b. An object passes through a cavity of another object.
   Example: Telescoping antenna

Table 4. Principle 4 — Asymmetry

a. Replace a symmetrical form with an asymmetrical form.
b. If an object is already asymmetrical, increase the degree of asymmetry.

Inventive Principles

Two principles have been selected at random for discussion. Principle 7 — Nesting, has many applications, from the fisherman's collapsible pole, to the carpenter's folding rule, to the cook's nested pots (Table 3). Applications of this principle resolve space requirement constraints.

In our age of symmetry, Principle 4 — Asymmetry is counterintuitive (Table 4). However, an example of productive asymmetry is the reduction in the sound of snow tires on dry pavement by the irregular spacing of treads. These variations are slight, but significant for the fact that they run contrary to the assumptions of the quality specialist looking for uniformity.

For aesthetic reasons, motor and generator mounts are often designed with symmetrical shapes, but since the machines rotate, the load on the mounts is actually asymmetrical. To reduce weight and conserve material, mounts for nonreversible units should be designed to support only the loads they must bear (Figure 1).

The 40 Principles have a remarkably broad range of application; they even work in business environments. These principles are so powerful that simply looking at the list often stimulates several new ideas. Appendix C provides explanations for each principle. To appreciate the power of the principles, select one at random and think of all the related applications you have actually experienced. Next, take any product and see which principle(s) is present and which principle(s) would cause an improvement.

Contradiction Table

From the 40,000 patents he studied, Altshuller selected several principles for each combination of engineering parameters. Not all of the combinations have recommended principles for conflict resolution because there were few or no patents which illustrated resolution of that particular contradiction. The rows in Appendix D display the parameters needing improvement. The columns contain the parameters that have been degraded as a result of

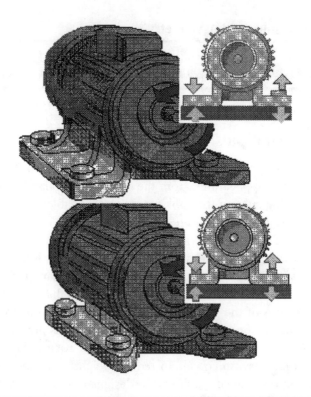

Figure 1. To reduce weight and conserve material, mounts for nonreversible motors and generators should be designed to support only the loads they must bear.

improving the parameter in the row. The recommended principles are found at the intersection of the rows and columns.

There is usually more than one principle that has been used in the past to solve any particular contradiction. In Appendix D, the principles are presented in order of decreasing frequency.

The screw problem must be structured to fit the generic parameters. Node 7 in Chapter 3, Figure 12 generated the problem statement, "***Find an alternative way to provide*** (screw integrated into bone), which does not require (large frictional forces)." This problem statement can be structured to fit the generic parameters used in the contradiction table. Bone growth around the screw can be considered an improvement in the parameter "Volume of a nonmoving object (8)," and the undesired high frictional forces needed to remove the screw can be defined as the parameter "Force (10)." The intersection of column and row suggests using Principles 2, 18 and 37 (Figure 2 and Table 5).

Figure 2. Contradiction table.

Feature to Improve \ Undesired Result (Degraded Feature)	1 Weight of moving object	2 Weight of non-moving object	* *	10 Force	* *	38 Level of automation	39 Productivity
7 Volume of moving object							
8 Volume of non-moving object				2, 18, 37			
* *							
* *							
39 Productivity							

Table 5. Recommended Principles

2. Extraction
18. Mechanical vibration
37. Thermal expansion

These generic concepts (defined in Appendix C) must now be adapted to the screw problem. The second (Mechanical vibration, Table 6) and third (Thermal expansion, Table 7) recommendations suggest promising principles.

Mechanical vibration suggests using the bone's resonant frequency. Small microvibrations, effective for loosening a rusted bolt, could work for the screw. Ultrasound could be directed to the thread area of the screw. The patient, however, felt that this proposal feels too risky, though piezovibration is intriguing. The principle of thermal expansion suggests cooling the screw to reduce its diameter.

Typically, several technical contradictions exist. Investigate all of them.

Table 6. 18 — Mechanical Vibration

a. Set an object into oscillation.
b. If oscillation exists, increase its frequency, even to ultrasonic levels.
c. Use the frequency of resonance.
d. Instead of mechanical vibrators, use piezovibrators.
e. Use ultrasonic vibrations in conjunction with an electromagnetic field.

Example

1. To remove a cast from the body without skin injury, a conventional hand saw is replaced with a vibrating knife.
2. Vibrate a casting mold while it is being filled to improve flow and structural properties.

Table 7. 37 — Thermal Expansion

a. Use expansion (contraction) of a material by heat (cold).
b. Use various materials with different coefficients of heat expansion.

Example

1. To control the opening of roof windows in a greenhouse, bi-metallic plates are connected to the windows. With a change of temperature, the plates bend and make the windows open or close.

Formulating the problem and choosing the direction in which to focus development are the most significant processes in innovation. The creativity of the team/individual is still needed to design an appropriate system to utilize the identified principles.

For the screw example, there are several technical questions which need to be answered for the concepts under consideration.

What is the resonant frequency for human bone, particularly the interior of the ball at the femoral neck? What is the coefficient of expansion/contraction for stainless steel and bone? These are engineering issues.

How much of a change in diameter is necessary to free the screw from the bone? How can the equipment be designed so that it will meet operating room requirements for sterilization? Addressing this latter requirement may involve using TRIZ again. These are also engineering issues.

Using the Contradiction Table: A Case Study

Consider an example of the piping necessary to pneumatically transport metal pellets. The existing system was originally designed to move plastic pellets, but a design change required metal pellets instead.

Before using the Contradiction Table, document the history of the problem. Metal piping controlled direction of the flow of plastic pellets; air flow moved the material; the same system moved metal pellets after a design change; downtime required for repairs became excessive; the metal balls quickly wore out the elbows in the system. Once this is recorded, several contradictions emerge within the history (Figure 3).

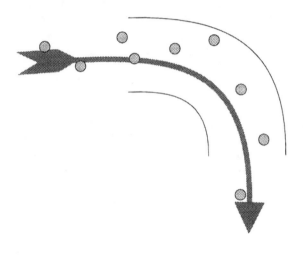

Figure 3. Elbow of piping to move pellets pneumatically.

Consider moving the material faster (increasing the speed) as an additional requirement. Conventional wisdom would reinforce the elbows required to change the direction of the pellets (Table 8). Two desired improvements include increasing the speed and decreasing the energy requirements (Table 9). There are several degrading parameters associated with each improvement.

There also are several parameters in this system that could be improved. The parameters which degrade when the speed of the pellets is increased are presented along with the principles used to reduce the contradiction. This information was taken from the intersections of the relevant parameters on the Contradiction Table. The principles are listed by decreasing frequency of occurrence (Table 10).

Table 8. Conventional Solutions

Reinforce elbows
Quick-change elbows
Redesign shape of elbow
Select another material for elbow

Table 9. What Is the Goal for the System?

Change direction of pellets
Reduce the energy requirements
Move the material quickly.

Table 10. Improving Speed (Parameter 9)

Degraded Parameters	Parameter Number	Principles Used
Reliability	27	11, 35, 27, 28
Force	10	13, 28, 15, 19
Durability	14	8, 3, 26, 14
Temperature	17	28, 30, 36, 2
Energy	19	8, 15, 35, 38
Loss of matter	23	10, 13, 28, 38
Quality of substance	26	10, 19, 29, 38
Harmful factors	31	2, 24, 35, 21

Table 11. Improving Energy (Parameter 19)

Degraded Parameters	Parameter Number	Principles Used
Convenient to use	32	28, 36, 30
Loss of time	25	15, 17, 13, 16

Table 12. The Frequency of Recommendations

Principle	Frequency	
28	5	(a) audio, optical, olfactory
		(b) magnetic, electric
		(c) replace fields
35	3	
13	3	
15	3	
38	3	

The parameters which degrade when the energy (Parameter 19) is decreased are presented along with the principles used to reduce the degradation (Table 11).

A tally of the principles used in all of the contradictions suggests looking at those that occur most frequently. The top five are listed in Table 12.

"Principle 28, Replacement of a mechanical system" is the most frequently recommended principle for all the contradictions considered. The existing elbow is a mechanical system.

"Replacement of a mechanical system" suggests:

a. replace a mechanical system with an optical, acoustical or olfactory system.
b. use an electrical, magnetic or electromagnetic field for interaction with the object.
c. replace fields:

   1. stationary fields to moving fields
   2. fixed fields to fields changing in time
   3. random fields to structured fields

d. use a field in conjunction with ferromagnetic particles.

An example is using an electromagnetic field to increase the bond of metal coating to a thermoplastic material. The process is carried out inside an electromagnetic field that applies force to the metal.

Alternative "b" suggests placing a magnet at the elbow to provide a blanket of pellets that absorb the energy. This idea triggered a lively discussion of various considerations, and electric charge on the balls, magnetizing the balls.

At times the introduction of a principle creates a secondary problem. If the secondary problem is easier to solve than the original primary problem, progress has been made. This line of reasoning will be used for all of the TRIZ tools presented in this text.

Physical Contradictions and Separation Principles

Physical contradictions are the mutually exclusive requirements of one state of existence along with an opposing state of existence. It also relates to a function, performance, or component (something must be slippery and rough). These contradictions are resolved by separating the requirements. The four main separation principles must be investigated because we do not know which will provide the best concept.

At times, the Technical Contradiction Table does not offer a convenient concept for resolving the contradiction. In these situations, the technical contradiction can be changed into a physical contradiction (Table 13).

Table 13. Physical Contradiction

Requires mutually exclusive states as they relate to a function, performance or a component.

The general process for changing a technical contradiction to a physical contradiction is to identify the characteristic of the desired and undesired result. This characteristic defines the physical contradiction. What is the physical contradiction for the technical contradiction? A good seal for the Top "A" of a glass ampule damages the medicine "B" in the container.

The team must consider what third function "X" is active to improve "A" but damage "B."

If the heat "X" improves "A," but degrades "B," then the physical contradiction is "X" is hot and "X" is cold (Table 14). There can be multiple physical contradictions, i.e., for the current damaging heat material "X" has a good seal but he medicine is damaged, but not using material "X" the seal is bad and the medicine is good. The contradiction is "X" must be used and "X" must not be used.

This process could transform a single technical contradiction into several physical contradictions.

Table 14. Technical and Physical Contradictions

Technical Contradiction
Heating ampule top "A" causes degradation to medicine "B."

Physical Contradiction
The heat "X" must be hot to seal and the heat "X" must be cool for the medicine.¹

---

1 Separation in space has bottom of the ampule in a cool liquid and the top receives the approriate heat.

Using our screw example, the team could reformulate a technical contradiction as a physical one.

During the formulation process for the screw in Chapter 3, Node 3 represented the statement, "(Large frictional force) is required for (screw integrated into bone) and causes [screwdriver slips in slot]." The technical contradiction is represented by, "Improvement in (screw integrated into bone) increases [screwdriver slips in slot]." The link is the influencing function "X" (large frictional force).

Having a large frictional force and *having a small frictional force* describe the physical contradiction. The *separation in time* principle suggests that the large frictional forces act while the screw is in use during bone growth, and low frictional forces act during removal of the screw.

The suggestion that the screw be removed using microvibrations at the resonant frequency of the bone was rejected by the patient because of a secondary problem. The technical contradiction of this recommendation for the patient is, "Improvement in (screw comes out easily) increases [bone is damaged]." The physical contradiction would be *high energy vibration* and *low energy vibration*. Again using the *separation in time* principle, the high energy could be used for a short burst and the low energy could be used during the turning of the screw. The change in frequency could possibly represent a tapping action.

If the screw problem was to design a new screw, then one of the problem statements for the formulation could be:

> Find an alternative way of "bind to the bone" which provides "holding the bone in position" and does not cause "is difficult to remove."

In normal English, "The screw must hold the bone in position and be easy to remove." The technical contradiction can be changed into a physical contradiction, "The screw is secure to the bone, the screw is not secure to the bone."

Using *separation in time*, we have, "The screw is secure during bone mending and comes out easily after the break is healed." A biodegradable screw is a possible solution, such devices are being investigated today.

Eyeglasses provide a simple example of three of the four commonly used separation principles. Eyeglasses are used to see near and far. If inventors had used the separation principles a century ago, they would have created today's designs a hundred years earlier. As a rule, each separation principle should be investigated because you do not know which one will lead to the most significant breakthrough (Table 15).

Table 15. Eyeglasses

Separation in Space: Two different lenses (bifocals).
Separation in Time: Two pairs of glasses, changing back and forth as the need arises.
Separation Upon Condition: The lenses are replaced with a self-focusing camera-type lens.

Table 16. Physical Contradiction Structures

A. Performing the function is necessary to achieve a desired result, and not performing the function is necessary to avoid harmful or undesired effects.
   Example: The pin of an IC chip must be heated to attach it to a circuit board and must not be heated to avoid damaging the chip.

B. A characteristic must be higher to achieve a desired result and must be lower to avoid harmful or undesired effects or to achieve another desired result.
   Example: An airplane wing should be large for take-off and small for high speeds.

C. An element must be present to achieve a desired result and must be absent to avoid harmful or undesired effects or to achieve another desired result.
   Example: Aircraft landing gear is necessary for landing but undesirable during flight.

If performing a useful function is associated with some harmful or undesired effect, decide which of the following contradictions (A, B or C) applies to your situation. There are three different ways to formulate physical contradictions (Table 16).

A. The first type of physical contradiction requires a function to be operational to achieve a desired result, but not operational to avoid harmful or undesired effects. For example, heat is applied to an IC chip pin that will be attached to a circuit board, but heat must not be applied because it damages the IC chip.
B. The second type of physical contradiction requires a characteristic to be of one value to achieve a desired result, but it must be the opposite to avoid harmful or undesired effects or to achieve another useful result. For example, an airplane wing should be large for take-off but small for high speeds. *Separation in time* is useful for this type of contradiction.
C. The third contradiction requires an element to be present to achieve a desired result but also absent to avoid harmful or undesired effects or to achieve another useful result. For example, aircraft landing gear is necessary for landing but undesirable during flight.

Four separation principles eliminate conflicts between requirements (Table 17).

Table 17. Resolution of Physical Contradictions

Separation in Space
Separation in Time
Separation Within a Whole and Its Parts
Separation Upon Condition(s)

Let's define each separation principle and explore an example for each one.

Separation in Space

The concept is to separate in space opposite requirements. If a system must perform contradictory functions or operate under contradictory conditions, try to (actually or theoretically) partition the system into subsystems. Then assign each contradictory function or condition to a different subsystem.

Example

Problem — Metallic surfaces are placed in metal salt solutions (nickel, cobalt, chromium) for chemical coating. During the reduction reaction, metal from the solution precipitates onto the product surface. The higher the temperature, the faster the process, but the solution decomposes at high temperatures. As much as 75% of the chemicals settle on the bottom and walls of the container. Adding stabilizers is not effective and conducting the process at a low temperature sharply decreases production.

Contradiction — The solution become apparent with a succinct rephrasing of the problem. The process must be hot (for fast, effective coating) and cold (to efficiently utilize the metallic salt solution). Using the separation principle in space, it is apparent that only the areas around the part must be hot.

Solution — The product is heated to a high temperature before it is immersed in a cool solution. In this case, the solution is hot where it is near the product, but cold elsewhere. One way to keep the product hot during coating is by applying an electric current for inductive heating during the coating process (Figure 4).

Separation In Time

The concept is to separate the opposite requirements in time. If a system or process must satisfy contradictory requirements, perform contradictory functions or operate under contradictory conditions, try to (actually or theoretically) schedule system operation in such a way that requirements, functions or operations that conflict take effect at different times.

Example

Problem — When an electrotechnical wire is manufactured, it passes through a liquid enamel bath and then through a die which removes excess enamel and sizes the wire. The die must be hot to ensure reliable calibration. If the wire feed is interrupted for several minutes or more, the enamel in the hot die bakes and firmly grips the wire. The process must then be halted to cut the wire and clean the die.

Contradiction — The die should be hot for operation and cold to avoid baking enamel. The separation in time principle suggests the die be hotwhen

Figure 4. Solution to the hot–cold problem.

the wire is being drawn and cold when the wire is not moving. Is there a way to have the die heated and not heated automatically? While the wire is being drawn on the die, there is a significant force pulling the die in the direction of the wire pull. When the wire stops there will be no pull.

Solution — The die can be fixed to a spring. When the wire moves, it pulls the die which compresses the spring into a heating zone. The die is heated either by induction or by contact with the hot chamber walls. When the wire stops moving, the spring pushes the die back into the cold zone (Figure 5).

At times, a contradiction is not initially stated as a time constraint. To resolve this contradiction, change the system's characteristics or environment.

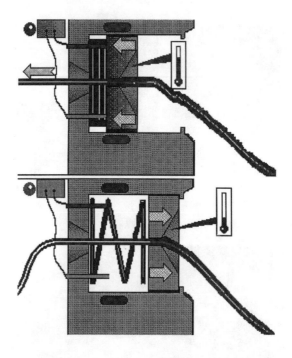

Figure 5. The solution to another hot–cold problem.

Example

Problem — When wide-screen movies first appeared, they were not broadly distributed because existing movie projectors in the majority of theaters could not accommodate the wider film. Distribution of the new format required the ability to use existing projectors to show wide-screen movies.

Contradiction — The film must be wide for the screen; the film must be narrow for the projector. The time contradiction was having one wide angle camera making the film and many traditional projectors showing the film months later. Starting with the latter condition, the traditional camera must have the wide angle view within a traditional frame.

Solution — One solution consisted of placing wide-screen frames lengthwise on narrow film by rotating the camera 90 degrees. Projector optics and mechanisms could be easily modified to accept the rotated frames (Figure 6).

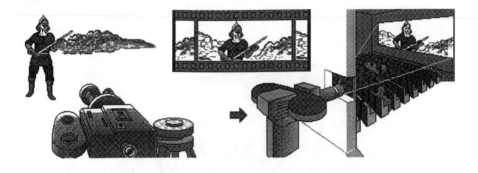

Figure 6. Solution to the wide-angle projector problem.

Another solution — Another solution is to optically compress the frame images so they fit in a conventional narrow-film frame. The frame images are optically expanded in the projector to provide wide-screen pictures.

The problem of placing the sound track still is to be resolved.

Separation Within a Whole Object and Its Parts

The concept is to separate the opposing requirements within a whole object and its parts.

If a system must perform contradictory functions or operate under contradictory conditions, try to partition the system and assign one of the contradictory functions or conditions to a subsystem (or several subsystems). Let the system as a whole retain the remaining functions and conditions.

Example

Problem — Work pieces having complex shapes can be difficult to grip using an ordinary vise.

Contradiction — The main function of the vise is to provide evenly distributed clamping force (a firm, flat grip face). The subsystem requires some means of conforming to the irregular shape of the object (a flexible grip face). The face must be flat; the face must be irregular.

Solution — Stand hard bushings on end between the flat surface of the vice jaws and the irregular surface. Each bushing is free to move horizontally to conform to the shape of the piece as pressure increases, while distributing even gripping force on the object (Figure 7).

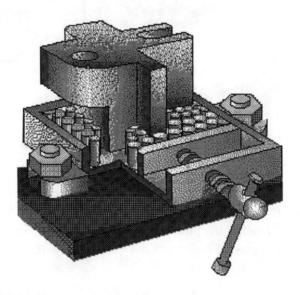

Figure 7. Solution to the vise problem.

Other Approaches to Separating Contradictions

An alternative for resolving a system with desirable and undesirable properties is to isolate the part(s) of the system or process that has the undesirable qualities.

Example

Problem — A soldering iron typically consists of a hollow shell which surrounds a heating element. This shell gets hot and can burn the operator.

Contradiction — Soldering equipment must be hot and cool.

Solution — If the space between the heating element and the shell is filled with heat-insulating foam, the danger of injury decreases. This is a Level 1 innovation because the solution was a simple insulation increase (Figure 8).

Another approach to resolving contradiction in a system with desirable and undesirable properties is to influence the offending component within

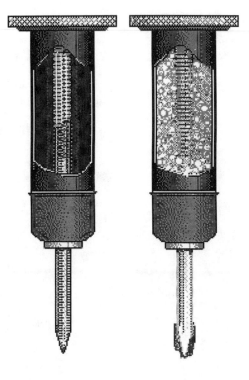

Figure 8. Making a soldering iron hot and cool.

the system (for example, isolate it or change its characteristics). Consider making use of the special properties or features of the component. Determine and apply a stimulus to initiate these properties or features.

Example

Problem — In steel casting operations, it is difficult to separate slag from molten metal.

Contradiction — Combine molten minerals to form an alloy and do not combine impurities from the minerals in the alloy.

Solution — A magnetic field is applied to the mold into which the liquid steel and slag is poured. The magnetic field does not affect the slag, which rises to the top where it is easily removed (Figure 9).

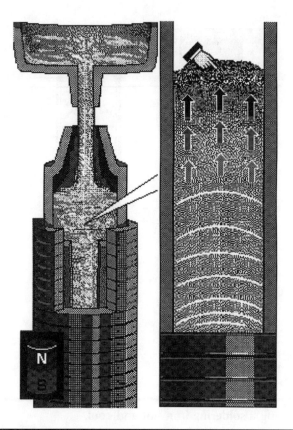

Figure 9. Separating slag from steel.

Separation upon Condition

The concept of separating opposing requirements of a condition(s) can resolve contradictions in which a helpful process takes place when special conditions exist. Consider changing the system or the environment so that only the helpful process can take place. In the kitchen, a sieve will stop the pasta but not the water.

Example

Problem — A lumber mill has a market for clean sawdust. A vacuum cleans the material around the saw blade, and the dust is collected via a sheet metal pipe.

Contradiction — We need a large pipe; we need a small pipe.

Solution — Increasing the diameter of the pipe for a short distance allows the heavier material to fall out.

A Word before Moving On

The shift in thinking needed to find the parameter that best describes your contradiction does not occur automatically. TRIZ practitioners must be persistent and disciplined in their use of the Contradiction Table. On the other hand, consider the time wasted trying to reach a solution through trial and error. Clearly, the benefits of this precise analysis of your problem speak for themselves.

The ideal design provides the desired function without a system. The next chapter discusses use of readily available resources as one way to reach the ideal design.

> "I have seen the future, and it works."
>
> — *Lincoln Steffens*
> *following a 1919 visit to the Soviet Union, in "Letters," 1938*

5 The Ideal Design

When to Use the Concept of Ideal Design

The gap between the current design and the *ideal system* should be reduced to zero. The ideal system provides the desired function without existing. This model becomes a goal to attain, shattering many traditional images of the most efficient system.

The ideal concept is a global concept, but the solution or approximation is dependent on local solutions. The resources are different for different individuals and in different locations.

The notion of identifying a primary useful function that is satisfied from a nonexistent system leads to design innovations within a very short period of time. Designers for lunar vehicle lights were having difficulty finding a covering for the bulbs which would withstand the vibrations and shocks of space travel. The problem was finally solved by proposing the lights used on tanks. When the lead scientist saw the design, he asked why they had worried about a cover to keep out oxygen when there was effectively no oxygen on the moon. The partial vacuum of the moon was a resource that eliminated the need for the cover. The function was satisfied without a system.

Very late in the space program to Venus, an influential scientist wanted his 10 kilogram experiment to be included on the voyage. He was told it already was too late because every gram already had been identified. Not accepting this answer, he identified 16 kilograms of ballast. He then proposed the removal of 10 kilograms to be replaced with his equipment. The ballast was an unidentified resource.

In other words, function is ideally performed by already existing resources. The concept of the ideal design should be consciously included during any application of TRIZ. Stating the ideal final result and backing away from it as little as possible offers a different technical challenge than the one offered by the technical contradiction.

A Case Approaching the Ideal Design

A simple example illustrates how the ideal final result is often performed using existing resources. A standard lightweight backpacking stove fueled by white gasoline works when white gasoline is in a gaseous state. To operate, a few drops of gasoline are placed in a depression in the gasoline tank. The small pool of gasoline is ignited around the brass pipe joining the tank to the burner. The hot air in the container creates pressure, driving the fuel up the tube. Once the liquid is preheated to a gas, the process becomes self-sustaining: the heat transfer moves from the cooking flame down the brass construction, where it forces the pressurized liquid through the hot brass structure. This interaction explains why winter campers who place the brass reservoir on snow or ice cannot get the stove to function. Some form of insulation is necessary between the stove and the snow. In this scenario, the heat loss that snow or ice causes is a negative environmental resource (Figure 1).

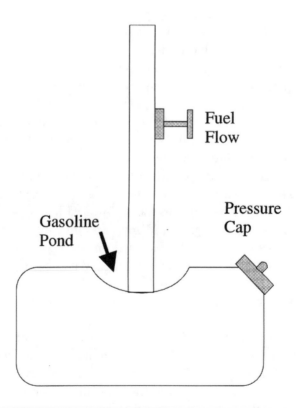

Figure 1. A typical backpacking stove.

Another application of environmental factors becoming a positive resource is a heat sink (snow) found in Sweden. A 1996 research project conducted in Sweden demonstrates how the environmental factors that performed a harmful function in the initial scenario act as positive resources in a different application. Researchers at the Royal Institute of Technology in Stockholm, Sweden (Anders Killander) studied alternative ways to generate electricity in rural areas far from utility lines.[24] The ideal final result would generate electricity from nothing.

Using analogous thinking, one may pursue the following reasoning. The primary difference between a motor and a generator is the reversal of input and output. A generator converts rotational energy into electricity, while a motor converts electricity into rotational energy. Measurement systems exist that depend on small changes in some physical property. Finding a physical phenomenon which uses or generates electricity may offer a concept for this research project.

Looking at the physical effects described in Appendix A, measuring temperature offers several approaches. The Seebeck phenomenon is one option. A thermoelectric generator based on the Seebeck effect is very promising. Anders Killander, the project's lead researcher, used TRIZ during this investigation. (Without the knowledge base contained in Appendix A, few electrical engineers would be familiar with the Seebeck effect.)

In 1821, T. J. Seebeck discovered that an electrical current is created in a closed circuit made of two conductors of heterogeneous metals, if the conductors' temperatures are different. The thermoelectric EMF (electromagnet force) which generates the current is directly proportional (in the first approximation) to the temperature difference between the two conductors. The proportionality coefficient is called the thermocouple constant, and it depends primarily on the contacting surface types (top of Figure 2). For metal thermocouples, the thermocouple constant is 10–50 microvolts/K; in a semiconducting thermocouple, the value is larger (for example, 0.1 V/K).

The Seebeck effect often is used to measure temperature and is widely exploited in devices that directly transform heat energy into electric energy. For metal thermocouples, the efficiency of transformation is about 0.1 percent; for semiconducting thermocouples it is 15 percent or more (as in the case of multicascade thermal batteries). To generate electricity, a device with no moving parts was placed on a wood burning stove. It included fins which were cooled to provide the temperature differential (bottom of Figure 2). This solution underscores how practitioners build their own cross-referenced library of definitions for effects. The current design using the Seebeck effect to supply limited power to rural homes is affordable. The unit costs about $150. There is concern that the current modest demands of the rural community will increase beyond the capabilities of the device as the convenience of electricity becomes part of daily life. This possibility provides the opportunity for a new TRIZ application.

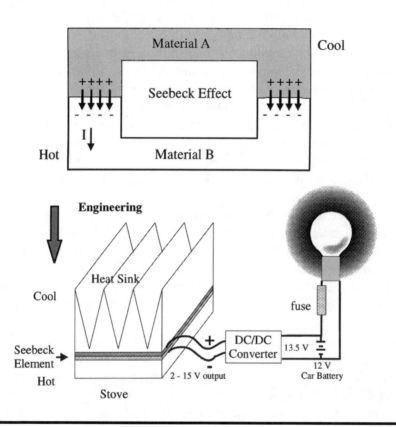

Figure 2. Applying the Seebeck effect to supply limited power to a rural home.

It is humbling and inspiring to access old or forgotten knowledge that has so many applications in our technological age. Many of today's electrical devices require very little power, which makes generators based on the Seebeck effect feasible. A limited knowledge base reduces our options. The following tools expand intellectual resources, opening up a world of innovative possibilities. The table in Appendix A was created by Altshuller in 1976.

A more comprehensive list is available in a Russian booklet by Gorin[25] and the Russian journal *Physical Effects for Inventors and Innovators.*" [26]

What Is Ideality?

Ideality is defined as the sum of a system's useful functions divided by the sum of its undesired effects.

$$Ideality = \frac{All\ Useful\ Effects}{All\ Harmful\ Effects}$$

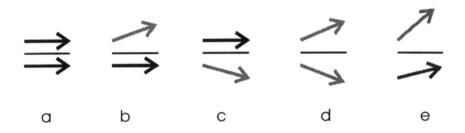

a   b   c   d   e

Any form of cost (including all types of waste and pollution) is included in the undesired effects. These system costs include the space it occupies, the noise it emits, the energy it consumes, etc. Design changes resulting in any combination of increases to the numerator and decreases to the denominator bring the system closer to the ideal design.

This theoretical measure suggests we look at designs that:

a. accept current design.
b. increase the numerator by adding functions or by improving the performance of some functions (the more important ones).
c. remove unnecessary functions in order to reduce the denominator.
d. combine the subsystems for several functions into a single system in order to decrease the denominator.
e. increase the numerator at a faster rate than the denominator.

An Ideal Container is No Container

One example of improving ideal design involves comparing the resistance of different alloys to an acid. Several alloy specimens are placed in a closed, acid-filled container. After a predetermined time, the container is opened. The effect of the acid on the specimen is measured. Unfortunately, the acid damages the container walls. The walls can be coated with glass or some other acid-resistant material, but this solution is costly. The Ideal design has a specimen exposed to the acid without requiring the use of a container. The transformed problem is to find a way to keep the acid in contact with the specimen without a container. Some of the resources available are the specimen, air, gravity, adhesion, etc. The solution is now obvious; make the container out of the specimen (Figure 3). Increating the number of specimens tested is a bonus, because the size of the container no longer is a limiting factor.

TRIZ provides two general approaches for achieving close-to-ideal solutions (which increase the ratio of useful functions divided by harmful functions).

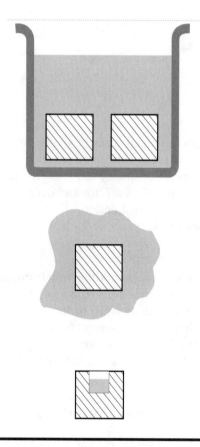

Figure 3.   Using the specimen as the container.

Use of Resources

A resource is any substance (including waste) available in the system or its environment that has the functional or technological ability to jointly perform additional functions. Some examples of resources are energy reserves, free time, unoccupied space, information, etc. Members of the TRIZ community jokingly conclude that "nothing" is the most useful resource. Humor aside, nothing is really something if it comes in the form of dead space for temperature and sound insulation. ThermoPane windows and sound baffles are two examples of nothing doing something.

Use of Effects: Physical, Chemical, Geometric, etc.

Often a complex system can be replaced with a simple one if a physical, chemical or geometric effect is used. For example, reinforcing rods are stretched before pouring concrete during the manufacture of prestressed concrete slabs. Instead of a hydraulic system, utilize the coefficient of thermal expansion by heating the rods, causing them to expand. The rods are then clamped into position and allowed to cool.

Let's take a closer look at how the use of resources improves designs. A North American company needed the ideal clip for hanging a repositionable easel pad. Their solution is startlingly simple; use the pad itself. Make each sheet repositionable, and use the sheet adhesive on the back of the pad. The pad can now be placed on an easel or directly on a wall.

Natural phenomena are relatively free resources. For example, during the colonial period in North America, workers split granite by placing water in drilled holes. The increase in volume that occurs when water undergoes a phase change from liquid to solid (freezing) fractured the granite.

Sometimes poorly conceived designs result from ignoring natural phenomena. An old heating chamber for curing plastic had hot air entering at the center of the floor and flowing around a complex shape before exiting through the top. A new $1 million chamber had hot air coming in from the top and exiting out the side. Organizations buying the new system had to do extensive research to produce a baffle configuration, recreating the former naturally occurring uniform temperature gradient. The older design took advantage of the natural phenomenon that warm air rises. More than 250 physical effects are available. (Thermal expansion can be used for for precision adjustment.) Furthermore, more than 120 chemical effects (etching to remove material) and 50 geometric effects (a Möbius strip to increase surface area) can be applied to a variety of applications. What problem solvers need is a convenient textual presentation of this information.

Six approaches to finding the ideal system are offered in Table 1.

Table 1. Finding Ideality

1. Exclude auxiliary functions.
2. Exclude elements.
3. Identify self-service.
4. Replace elements, parts, or total system.
5. Change the principle of operation.
6. Utilize resources.

Six Paths to Improve Ideality

1. Exclude auxiliary functions

Auxiliary functions provide support for and/or contribute to the execution of the main function(s). In many situations, auxiliary functions may be excluded (together with the elements and/or parts associated with their performance) without deteriorating the performance of the main function(s).

Example: Painting without solvents.

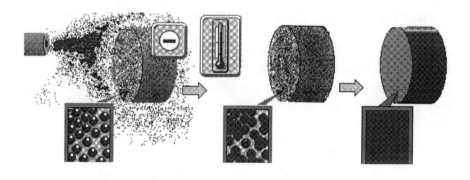

Figure 4. Painting without solvents.

Painting metal parts with conventional paint releases dangerous fumes from the paint solvents. An electrostatic field can be used to coat metal parts with powdered paint. After the powder is applied, the part is heated and the powder melts. A finished coat of paint is thus produced without solvent (Figure 4).

2. Exclude elements

Consider excluding system elements by delegating their functions to resources. For this purpose, use the available resources identified in ISQ of Chapter 2.

Substance Resources

Substance resources include any materials of which the system and its surroundings are composed. Readily available resources include: waste, derived resources and substance modification.

Waste
   Raw materials or products
   System elements
   Inexpensive substances
   Water

Derived resources
   Transformed waste
   Transformed raw materials or products
   Other transformed substances
   Modified water

Substance modification
   Phase transformations
   Chemical reactions
   Application of physical effects
   Heat treatment
   Fractions
   Decomposition
   Transformation to a mobile state
   Formation of mixtures
   Introduction of additives
   Ionization (recombination)
   Physical/chemical water treatment

Example (substance modification — ionization): Improving chemical welding accuracy by using a physical effect.

Workpieces that cannot tolerate high temperatures can be joined together by chemical welding. A reagent that reacts with both workpieces is used to form the desired weld.

To improve the accuracy of chemical welding, a reagent that reacts when exposed to ultraviolet light can be used. When the workpieces and the welding reagent are properly positioned, a beam of ultraviolet light is focused on the weld area and the welding action takes place (Figure 5).

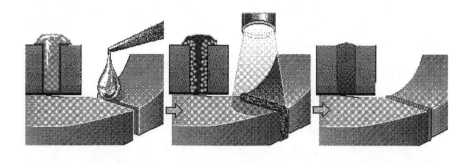

Figure 5. Ultraviolet light used in chemical welding: (1) Applying the reagent, (2) exposing the reagent to the ultraviolet light, and (3) the finished weld.

Functional Resources

Functional resources include the capability of a system or its surroundings to perform additional functions. A super effect is an additional (unexpected) benefit that arises as a result of innovation. For example, by exercising one arm it is expected that bone mass will increase. The super effect is that, to a lesser extent, the bone mass in the other arm also increases. Resources include:

Application of existing functions
Application of super effect
Application of harmful functions.

Example (Functional resource — application of harmful functions): Making use of blood incompatibility.

If blood of an incompatible type is used in a transfusion, clots will form when the transfused blood mixes with the patient's blood. This effect can be used in an emergency to help stop bleeding by applying a pad saturated with incompatible blood to a wound (Figure 6). With today's AIDS problem, this solution could have a harmful effect.

The Ideal Design

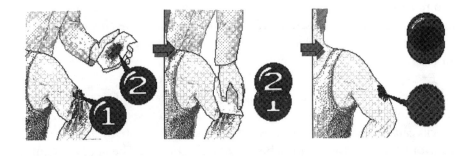

Figure 6. A pad saturated with incompatible blood could cause a wound to stop bleeding causing the flowing blood to clot.

Field Resources

Field resources can replace system elements. The difference in electric potential between the ionosphere and the ground produces a low-capacitance electric field of approximately 100 V/m. (This means that the voltage between the feet and head of a human, while outdoors, is about equal to his or her height in centimeters.) It has been suggested that this electric field be used to control airplanes flying at a low altitude (Figure 7).

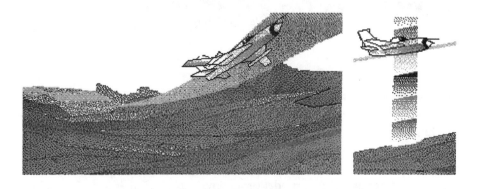

Figure 7. A low-capacitance electric field could be used to control airplanes flying at a low altitude.

3. *Identify self-service*

Test your object for self-service. Look for auxiliary functions fulfilled simultaneously with or at the expense of primary useful functions. You may also replace the means to fulfill the auxiliary functions with the means used for the primary useful function. The system becomes more efficient without corresponding auxiliary elements. The corrosion testing presented earlier is an example of self-service (Figure 3).

Example: Turning heavy rotors during transport.

Electrical machines, turbojet engines and other machinery with heavy rotors mounted on ball bearings are difficult to transport: vibrations and shocks cause the stationary bearings to produce hollows on the bearing races. To avoid this, the rotors must be periodically turned. The special engines required to do this are expensive.

The shocks themselves can be used to turn the rotor. A load in the form of a pendulum is attached to the rotor shaft and connected to a ratchet that allows movement in only one direction. The force from the shock causes the pendulum to swing up and, as it returns, turns the ratchet, thus forcing the rotor to turn (Figure 8).

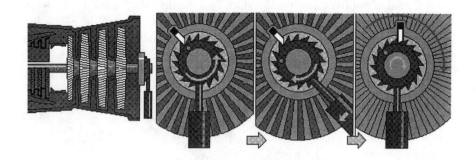

Figure 8. A pendulum can be used to turn rotors during transport of heavy machinery.

4. *Replace elements, parts or total system*
    Consider applying a model or a copy.
    Consider replacing a complex product (or part of it) with a simplified version or a copy.
    Consider using (temporarily or permanently) a copy or image of the object.

Consider using a life-size or scale model from which the elements responsible for an undesired property can be eliminated or replaced. In particular, consider the use of simulation.

Example: Simulating landing wheel tire traction.

The traction of aircraft tires during landing is uncertain in rainy weather. To get up-to-the-minute data on landing gear traction, a test vehicle can be fitted with a wheel that simulates the operation of a landing-gear wheel. This test wheel rotates at 90% of the speed of the other wheels. As the test car moves across the runway, a portable computer processes transducer signals from the test wheel. Then the results are radioed to landing planes. Airports in Washington, Hartford, Buffalo, Detroit, and Atlanta employ this system (Figure 9).

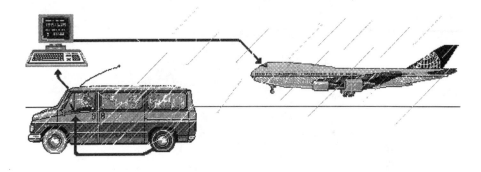

Figure 9. A test vehicle can provide current information on runway traction to incoming aircraft.

5. *Change the principle of operation*

To simplify a system or process, consider changing the basic operating principles.

Example: Keeping soft-sheet glass flat.

Hot, soft-sheet glass (used to manufacture plate glass) tends to sag between the rollers as the sheet moves on a conveyor. The ideal system has no sagging. If the rollers were smaller, the sag would decrease. What is the smallest roller? A molecule. A TRIZ solution is to convey the hot sheet and keep it flat, floating it on a pool of molten tin (Figure 10).

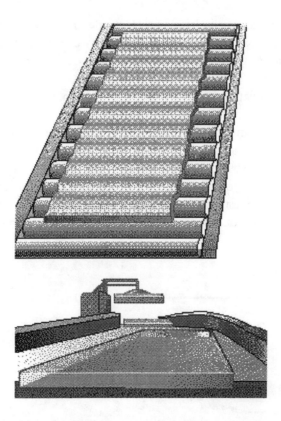

Figure 10. Floating hot sheet glass on a pool of molten tin (bottom) prevents it from sagging (rippling) during manufacture, as it did when rollers were used (top).

6. *Utilize resources*

Resources are substances, fields (energy), field properties, functional characteristics and other attributes existing in a system and its surroundings, which are available for system improvement.

Resources can be divided into several categories. Readily available resources are resources that can be used in their existing state. Derived resources are resources that can be used after some kind of transformation. Substance resources, field resources, functional resources, space resources and time resources are all elements available to most systems.

Substance Resources

Since substance resources include all material from which the system and its surroundings are composed, any system that has not reached the ideal should have substance resources available.

Example: Using a pollutant (waste) to prevent pollution.

To prevent pollution, exhaust gas from thermal power stations is treated with alkaline chemicals. The alkaline slag is itself recovered from coal burning coal power stations, where the slag had also been a source of pollution (Figure 11).

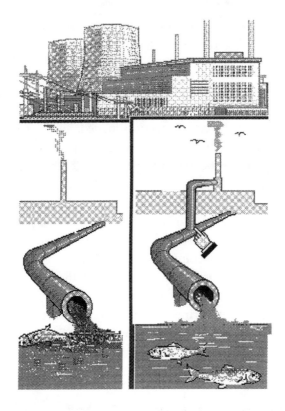

Figure 11. Alkaline wastewater can be used to treat exhaust gases, each neutralizing the other.

By using the alkaline wastewater (from cleaning the slag) to treat the exhaust gases, two harmful effects are used to neutralize each other.

Example: Using a product (system element) to prevent pollution.

Exhaust gases from excavators and heavy trucks used in deep, open pits are filtered using the extracted product (for example, crushed rock, coal, sand, etc.).

Figure 12. A truck's load, such as quarried material, can be used to accumulate harmful exhaust fumes.

Truck exhaust pipes are directed into the truck bed, where exhaust fumes are refined as the harmful contents accumulate on the rock (Figure 12). This method only can be used when the truck is loaded; however, an empty truck produces less exhaust. Other possible solutions can be used, depending on the available resources in the open pit. For example, snow is used as a filter in northern regions.

Example: Using a system element to measure temperature.

To prevent overheating of machine components (for example, bearings), a temperature control system is installed that typically includes thermocouples positioned where overheating is likely to occur.

Sliding bearings often include an electroconductive insert in an iron ring fixed within the component body. Overheating can be prevented by using the contact between the iron ring and the body as a thermocouple. That is, the component is switched off if the thermocouple detects a value above a certain temperature (Figure 13).

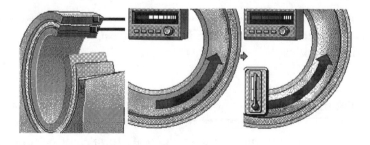

Figure 13. Sliding bearings can be modified to connect to a thermocouple to prevent overheating.

Example: Using a natural resource to prevent hazard.

One of the dangers in a coal mine is the possibility of coal dust explosions. In northern coal pits, snow is blown into the area to prevent these explosions. The snow accumulates and slowly melts, cooling the air as it does so (Figure 14).

Figure 14. Snow can be used to prevent coal dust explosions by cooling the air in a coal mine.

Example: Using a natural resource to measure an attribute.

During the manufacture of industrial ceramic vessels having irregular form and narrow necks, the wall thickness of the vessels must be measured. Water can be used for this measurement (Figure 15).

Figure 15. Water and an ohmmeter can be used to measure the thickness of a vessel wall.

To accomplish this, the vessel is filled with water that has increased conductance due to the addition of salt. One electrode of an ohmmeter (a device which measures electrical resistance) is immersed in the water; the other electrode contacts the external surface of the vessel. In this way, a measurement of the resistance proportional to the thickness of the vessel wall is obtained.

Derived Resources

Derived resources are resources that can be used after some kind of transformation. Raw materials, products, waste and other elements of the system, including water, air, etc., that may not be useful in their existing state, might be transformed or modified to become such a resource.

Example: Modify waste to conserve resources.

Restaurants, bars and cafes use large quantities of soap for washing dishes. To conserve soap, the utensils can be soaked in sodium bicarbonate before washing. Bits of fat on the utensils react with the bicarbonate, forming salts of fatty acids — in other words, soap. A soapy film now covers the utensils most in need of cleaning, and less soap is necessary (Figure 16).

Some greenhouses store heat in water containers or rocks. Water only stores one calorie per degree of Celsius for each gram. But without any change

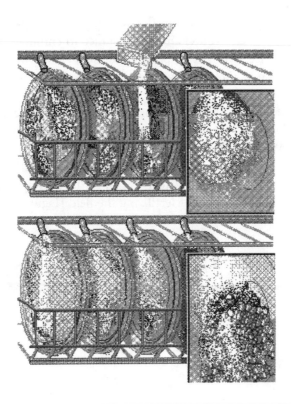

Figure 16. Sodium bicarbonate may be used to presoak dishes, forming soap when it combines with the fatty acids of food residues.

in temperature, each gram of ice requires 79.9 calories for the phase change from solid to liquid.

The Handbook of Chemistry and Physics (CRC Press) lists compounds which have a phase change at 26°C. The closest is *tert*-Butyl alcohol at 25.4°C with a heat of fusion at 21 calories.[27] Thus a small volume can store more energy, which can be released at a critical temperature by taking advantage of a phase change.

Substance Modification

As with derived resources, substance modification can be used to overcome apparent obstacles. Look for ways to overcome an obstacle by changing something within the existing system.

Can you change a substance in the system to create space, time or the necessary object, or can you change a substance to eliminate an undesired object? For instance, it might be possible to facilitate a removal process by changing the state of the object you want to remove through sublimation, evaporation, drying, grinding, melting or dissolving.

Example: Melt an object after it has served its useful function.

Clay disks launched for moving-target practice (skeet shooting) are called clay pigeons. When clay pigeons are used in skeet shooting, the ground becomes littered with clay fragments. However, disks made of ice are less expensive, and fragments falling to the ground melt away. Disks made of dung provide a new function of fertilizing the field (Figure 17).

Figure 17. **Clay pigeons made of ice or dung prevent the buildup of clay in skeet fields.**

Time Resources

Time resources include time intervals before the start, after the finish, and between cycles of a technological process which are partially or completely unused. Find time resources by:

- alteration of an object's preliminary placement.
- application of pauses.
- use of concurrent operations.
- elimination of idling motion.

Example: Give an object adjustable orientation so that it can work while moving in two direction, instead of only one.

In agriculture, each row is normally plowed in the same direction so that the upturned soil always lies on the same side of the furrow. To start each new row, the plow must make an idle return run.

Using a plow with both right-hand and left-hand blades saves time. The operator presses a button to switch blades at the end of a row, and then immediately plows the next row. The soil still lies on the same side of each furrow (Figure 18).

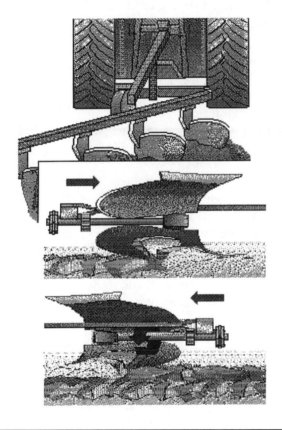

Figure 18. A plow with both right- and left-hand blades allows the farmer to plow continuously

The plow example not only shows the use of derived resources, but also demonstrates cyclical evolution; that is, the reuse of old concepts in more sophisticated systems. The first metal plows were designed to be pulled by draft horses, and the blade turned the soil in one direction. A later model had the advantage of a blade that could rotate in either direction, though it did not turn the soil as well. Initially, tractors went back to plowing in one direction. One compromise for time efficiency, though it added extra weight, was to carry two plows, one for turning the soil to the left, one for turning it to the right. Today's tractors use a hydraulic system to change heavy blades, while horse-drawn equipment required only the farmer's foot.

Application: Using Four Steps for Ideal Design

Step 1. Describe the situation you would like to improve.

Because of the intense heat in a furnace, the walls are cooled with water. The cooling system uses water pumped through pipes. If a pipe cracks, then water leaks out. This can cause an explosion in the furnace.

Step 2. Describe the ideal situation.

Water remains in the pipes even when there is a crack. Stated more aggressively: The water does not want to leave the pipe.

Step 3. Can you think of how the ideal situation might be realized? In other words, is there a known way to realize it?

If "yes": Congratulations! You have an idea! Be sure to document it.

If "no": Consider how to utilize available resources.

If "yes," but doing so is associated with some drawback: Go to resolving a contradiction.

If there is an obstacle that prevents you from realizing the ideal situation, describe what it is and why it is an obstacle:

"The pressure inside the pipe is greater than the pressure outside the pipe."

Step 4. Do you know what change(s) should be made to overcome the obstacle?

The pressure inside the pipe should be made lower than the pressure outside. Therefore, a vacuum water pump should be used.

In the next chapter, substance–field analysis is directed at investigating the impact of different energy fields on design. This tool provides another perspective for improving the innovative paradigm.

"Don't fight forces; use them."

—R. Buckminister Fuller

6 System Modeling, Substance–Field Analysis

The Substance–Field (Su–field) Model

Substance–field (Su–field) analysis is a TRIZ analytical tool for modeling problems related to existing technological systems. Every system is created to perform some functions. The desired function is the output from an object or substance (S1), caused by another object (S2) with the help of some means (type of energy, F). The general term, *substance* has been used in the classical TRIZ literature to refer to some object. Substances are objects of any level of complexity. They can be single items or complex systems. The action or means of accomplishing the action is called a *field*. Su–field analysis provides a fast, simple model to use for considering different ideas drawn from the knowledge base.

Su–field analysis works the best for well-formulated problems, like those developed with the formulation process or structured as a contradiction. Also, this analytic instrument requires greater technical knowledge (information on how to perform the physical effects for example) than some of the other TRIZ tools.

When to Use Substance–Field Analysis

Two substances and a field are necessary and sufficient to define a working technical system (Figure 1). The formation of this trilogy can be found in the early work of the mathematician Ouspensky.[28] The triangle is the smallest building block for trigonometry, as well as for technology.

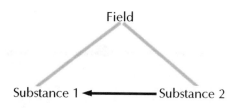

Figure 1. The triangle is the smallest building block.

There are four basic models:

1. Incomplete system (requires completion or a new system).
2. Effective complete system.
3. Ineffective complete system (requires improvement to create the desired effect).
4. Harmful complete system (requires elimination of the negative effect).

If any of the three elements are missing, Su–field analysis indicates where the model requires completion and offers directions for innovative thinking.

If the three required elements, Su–field analysis can suggest ways to modify the system for better performance. This is particularly true if radical changes in the design are possible.

Following the analogous thinking of TRIZ, a triangular technical system should have its own set of rules within the geometry of problem-solving. These few basic rules and the 76 Standard Solutions permit the quick modeling of simple structures for Su–field analysis.[29]

Making a Model

The field, which is itself often some form of energy, provides some energy, force or reaction to guarantee an effect. The effect could be on S1 or the output of the field information. The term field is used in the broadest sense, including the fields of physics (that is, electromagnetism, gravity and strong or weak nuclear interactions). Other fields could be thermal, chemical, mechanical, acoustic, light, etc.

The two substances can be whole systems, subsystems or single objects. They can also be classified as tools or articles. A complete model is a triad of two substances and a field.

The innovative problem is modeled as a triangle to illustrate the relationships between the two substances and the field. Complex systems can be modeled by multiple, connected Su–field triangles.

There are four steps to follow in making the Su–field model:

1. Identify the elements.
   The field is either acting upon both substances or is integrated with substance 2 as a system.
2. Construct the model.
   After completing these two steps, stop to evaluate the completeness and effectiveness of the system. If some element is missing, try to identify what it is.
3. Consider solutions from the 76 Standard Solutions.
4. Develop a concept to support the solution.

In following Steps 3 and 4, activity shifts to the other knowledge-based tools.

A flow chart showing how the problem solver would apply this TRIZ tool is seen in Figure 2. You can see that there is a constant alternation between the analytic and the knowledge-based tool.

The process cycles within Steps 1 and 2 until a complete model is found. The standard solutions in Step 3 offer breakthroughs in thinking. Alternative structures to the complete system are considered. For each structure, alternatives to the basic building blocks are considered through the use of knowledge-based tools.

Su–field analysis was created during 1974–1977. Each cycle of improvement increases the number of recommendations available. Today, there are 76 Standard Solutions. The 76 Standard Solutions are refinements of the original solutions. Only four models are presented in this material.

Breaking of a rock with a hammer is often used to introduce Su–field analysis. A modification of this simple example will be used to present some of the standard solutions.

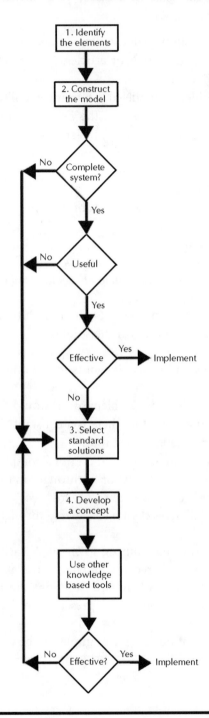

Figure 2. A TRIZ-oriented flow chart for problem solving.

Analysis Nomenclature

The identification of substances (S1 and S2) depends upon the application. Either substance could be a material, tool, part, person or environment. S1 is the recipient of the systems action. S2 is the means by which some source of energy is applied to S1.

The source of energy, or field (F), which acts upon the substances often is:

    Me — mechanical
    Th — thermal
    Ch — chemical
    E — electrical
    M — magnetic
    G — gravitational

The letter(s) associated with the applied field will be used in the triangular model of the different systems.

Relationships between the elements in the Su–field model are depicted by five different connecting lines:

1. Application
2. Desired effect
3. Insufficient desired effect
4. Harmful effect
5. Transformation of model

Analysis

Apply the four modeling steps to the four basic models.

    1. *Identify the elements*
   Our task is to break a rock.
     Function = Break Rock
     Rock = S1
   The system is missing the tool and energy source.
     Tool = S2
     Energy Source = F

2. Construct the model

Incomplete system — The rock is S1. If there is only the rock, it will not break, and therefore the model is incomplete (Model a, Figure 3). The model also is incomplete if only the rock and a hammer (S2) exist (Model b, Figure 3). In the same way, the model is incomplete if some field (gravity) and the rock are the only elements of the system (Model c, Figure 3).

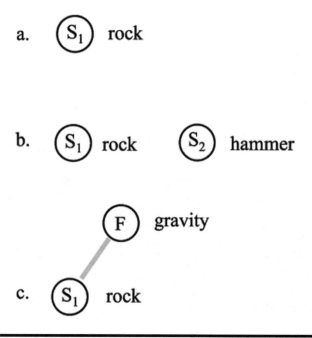

Figure 3. Incomplete models.

In these incomplete models, the desired effect does not occur. Completing the system, at the very least, will make the useful function possible. One complete system is a pneumatic hammer which provides mechanical force upon the rock with the hammer. The performance of the system is not considered at this time. The incomplete Model b is completed by the application of the mechanical field (F_{Me}) by the hammer (S2) to the rock (S1), as shown in Figure 4.

Once a complete system has been defined, performance can be analyzed. The evaluation of a completed system's performance gives three possible answers — an effective complete system, a harmful complete system, an ineffective complete system .

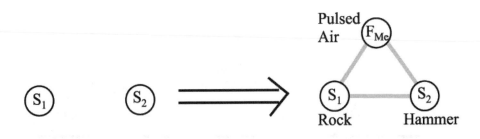

Figure 4. A model incorporating the elements to perform the desired task.

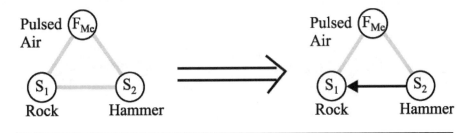

Figure 5. A complete system performing the desired task.

Effective complete system — If the system provides the desired effect, analysis is complete (Figure 5).

There are two ways in which the system might not provide the desired effect:

i. a harmful effect occurs.
ii. the results are inadequate.

3. Consider solutions from the Set of Standard Solutions

Harmful complete system — Of the 76 Standard Solutions,[29] many can be applied to resolve a harmful effect (Figure 6). Two ways of applying a standard solution are: introduction of another substance (Figure 7) or introduction of another field (Figure 8). Insights come from considering different substances for S and different fields for F.

Ineffective complete system — Standard solutions also can be applied to improve ineffective performance (Figure 9). As many alternative means as possible for new fields and substances should be considered. By adding to or changing the elements of the model, performance can be improved in six different ways. For example, change the substance (Figure 10). Or, change

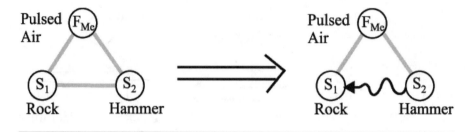

Figure 6. A harmful effect.

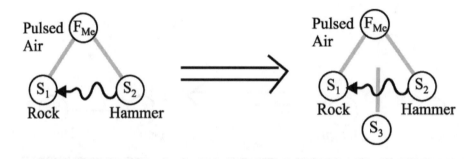

Figure 7. Introduction of another substance.

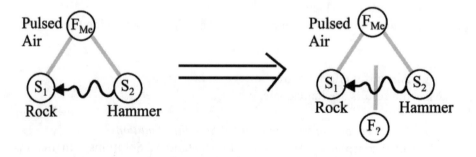

Figure 8. Introduction of another field.

the field and substance to a different mechanical field and a different hammer (Figure 11).

An additional field could be placed between the rock and the hammer (Figure 12). A chemical field to embrittle the rock would be effective.

An additional substance, or another field and substance could also be added (Figure 13).

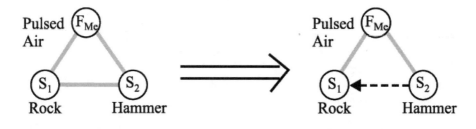

Figure 9. Applying a standard solution to improve ineffective performance.

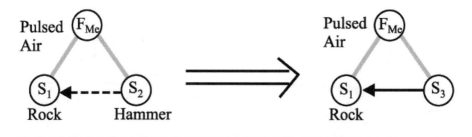

Figure 10. Improving performance by adding to or changing elements of the model.

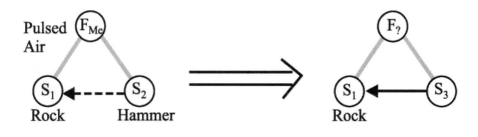

Figure 11. Changing the field and substance to improve performance.

Each of these solutions can lead to several new design concepts. The 76 Standard Solutions only provide a structural change to the model. The innovative problem solver must still develop a concept to support the structural change.

4. *Develop a concept to support the solution*

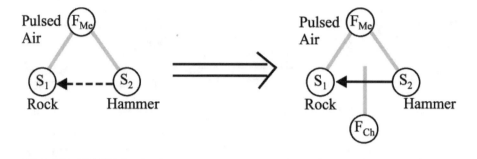

Figure 12. Improving performance by using an additional field.

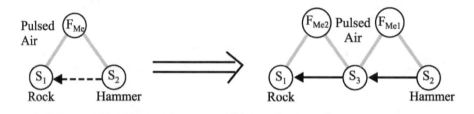

Figure 13. Improving peformance by adding an additional substance or another field and substance.

The structural changes identified in Step 3 direct the search for a means to support the change. Because the changes made are based only on the original form of the problem, new directions to search for design solutions are presented. Some of the directions may not be fruitful, but the important function of the model is to create global concepts. The challenging task is to supply detail to the global concept. Other TRIZ tools can be used to support this search.

Harmful complete system — If the harmful effect in our example is flying pieces, a metal cap or wire mesh covering the rock could be an added substance (S_3) used to eliminate the harmful effect (Figure 7). Look at all the fields that are available when adding a field (Figure 8) to a system. If the rock contains moisture, freezing (F_{Th}) could cause cracks produced by the expansion of the moisture when it freezes. The breaking will occur as these cracks gradually expand, reducing the explosive release of fragments. This result may also be considered a "super effect," as it reduces the mechanical effort required to achieve the function.

Ineffective complete system — Breaking of the rock may not be efficient or as effective as desired (Figure 9). One possibility for changing the substance (S3) in Figure 10 is to replace the original hammer head with a rock hammer head (S3). One way to change the field and substance in Figure 11 is to use a gas-fired thermal field (F_{Th}) and water (S3) to produce steam. The rapid temperature change could crack the rock. The added field in Figure 12 may be chemical (F_{Ch}), to make the rock (S1) more brittle. To add a substance and a field in Figure 13, a chisel (S3) can be placed between the hammer and the rock. There are now two systems with three elements. The air pressure (F_{Me1}) acts on the hammer (S2), transferring energy to the chisel (S3). The hammer provides the energy (F_{Me2}) to the chisel (S3), which transfers the energy to the rock (S1).

The old New England method of splitting rocks was to drill holes in the stone and place water in the holes during the winter. This model also would have two triads: first the mechanical field to drill the hole in the stone, followed by the thermal field applied by the water to the stone.

A Case Study

In the electrolytic processes for producing pure copper, a small amount of electrolyte remains in surface pores. During storage, the electrolyte evaporates and creates oxide spots. These spots result in a significant financial loss because of the defective appearance of the sheets. To reduce these losses, the sheets are washed before being placed in storage, but it is still difficult to remove all the electrolyte because of the small size of the pores. How can the washing process be improved?

1. *Identify the elements.*
   Electrolyte = S1
   Water = S2
   Mechanical process of washing = F_{me}

2. *Construct the model.*
In this case we have an insufficient desired effect because of the discoloration of the surface (Figure 14).

3. *Select a solution from the Standard Solutions.*
Adding a field to intensify the effect (of washing) is one standard solution (Figure 15).

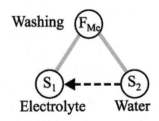

Figure 14. An insufficient desired effect.

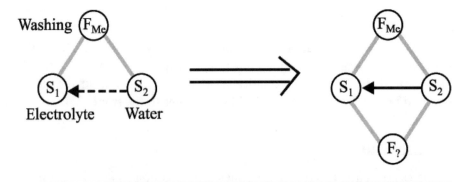

Figure 15. Adding a field to intensify the effect is one standard solution.

4. *Develop a concept to support the solution.*

There are several possibilities for fields which will intensify the effect of washing:

- a mechanical field using ultrasound.
- a thermal field using hot water.
- a chemical field using surfactants to dissolve the electrolyte.
- a magnetic field to magnetize water to improve washing.

Repeat the process in Step 3 by considering another standard solution. For each standard solution identified in Step 3, the related supporting concept is developed in Step 4. Explore all the possibilities. Ask every, "What?"

3. *Selecting a different solution from the Standard Solutions.*
Insert a substance S3 and another field F2 (Figure 16).

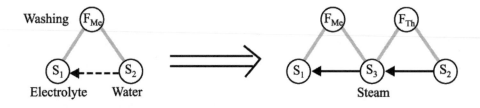

Figure 16. A solution different from the Standard Solutions.

4. *Develop a concept to support the solution*

F_{Th} is pressure and S3 is steam (Figure 16). Use superheated steam (water under pressure is hotter than 100°C). The steam will be forced into the pores, causing the electrolyte to leave.

Breaking a large problem into smaller pieces is a common process. Su–field analysis can be used first at the macrolevel, as well as at the microlevel. Every field option is another way of shattering the rock of psychological inertia that obstructs progress.

Identification of actual concepts that support proposed changes to a system is the work of the knowledge-based tools. Su–field analysis is structured innovation doing the challenging work of concept identification. The chapters presenting the knowledge-based tools support this activity.

Identifying the next generation of working systems offers a different type of design challenge. Insights into these challenges are offered in the next chapter, which explains standard patterns of evolution.

> *"Not that the story need be long but it will take a long while to make it short."*
>
> — *Henry David Thoreau*

7 Patterns of Evolution

When to Use Patterns of Evolution

If your goal is to gain a competitive advantage with a new design that is a quantum improvement over the current product, then the knowledge contained in the patterns of technological evolution is your most effective tool. The patterns are generic enough that they are also valuable in nontechnical applications. This chapter explores eight very general patterns of evolution. Each pattern contains its own detailed lines of evolution.

Solving a problem is reactive behavior meant to fix an ailing system. Looking at the future with patterns of evolution is proactive behavior designed to create the future.

Patterns of Evolution

A museum exhibit featuring several generations of designs for a single product offers a glimpse of evolutionary themes that are common to many products. These common themes of product/technology evolution provide a window into the future of other products. Identify the current position of today's design within an evolutionary pattern and you can predict future designs along this pattern. Thus, understanding the following eight patterns makes it possible to design tomorrow's products today. These eight patterns are listed in Table 1.

The safest approach is to plot the current design's development using several patterns of evolution; then look beyond today's position and consider the likely design evolution along each pattern. Combinations of evolutionary patterns are also informative. To facilitate this analysis, a matrix can be used in which both the rows and columns contain the patterns. The intersections represent new patterns. A crude cluster analysis of all the entries in the

Table 1. Patterns of Evolution

1. *Evolution in Stages*
2. *Evolution Toward Increased Ideality*
3. *Non-uniform Development of System Elements*
4. *Evolution Toward Increased Dynamism and Controllability*
5. *Increased Complexity then Simplification (Reduction)*
6. *Evolution with Matching and Mismatching Components*
7. *Evolution Toward Micro-level and Increased Use of Fields*
8. *Evolution Toward Decreased Human Involvement*

intersections will identify the more frequently occurring themes in future designs. The larger clusters indicate directions for development.

1. Evolution in Stages

The first pattern is the most generic, for it actually describes all evolution at the macrolevel. It is the common S-curve describing system performance as a function of time, represented by the top curve in Figure 1.

For many applications, the S-curve has a typical life cycle of pregnancy, birth, childhood, adolescence, maturity, and decline. Pregnancy is the time between an idea's inception and when it matures enough (or the environment is ready) for the public announcement of its birth. Birth marks the day a concept has clear definition and performs some function. Without development effort, a concept will not grow and evolve into a "mature" product. Many organizations view simultaneous engineering as a way to reduce development time. The biggest delay is often the time between an idea's birth and the moment when the idea becomes a project. Research-driven organizations can have ideas that wait 15 or 20 years (pregnancy) for a project, when development really starts. The S-curve is applied once an earnest effort towards development begins.

Three other curves plotted against time describe stages of evolution from a different perspective. The four curves in Figure 1 are aligned for comparison along the evolution timeline. Their various y-axes describe:

   a. performance
   b. level of inventiveness
   c. number of inventions (relating to the system)
   d. profitability.

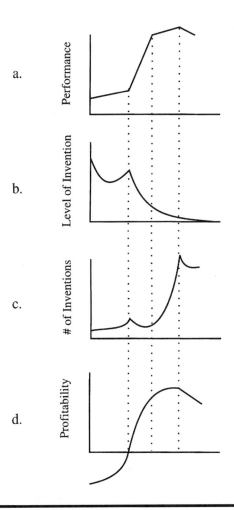

Figure 1. The common S-curve describes system performance as a function of time.

The initial concept is often very inventive and patentable, at Levels 3, 4 and sometimes 5 (Figure 1b). The difficult work that follows only produces improvements of the initial idea. Another breakthrough is necessary before the product is ready for the public, followed again by increased improvement.

While working toward this "breakthrough" innovation, the number of inventions per unit of time (that is, time needed to find solutions) increases until the breakthrough occurs (Figure 1c). Then the organization takes a brief rest before beginning the search for significant improvements. During the maturity of a product, many design improvements are necessary for marginal enhancement.

Obviously, profitability (Figure 1d) is negative during the start-up phase and peaks during maturity. Profit is realized when a substantial number of units are being produced. It is clear from these curves that a new design must be conceived during the rapid growth of the current design in order to maintain stable profit. Organizations frequently find that as an existing design is in its mature phase, they lack a new product or design approaching the rapid growth phase. Thus, profits may drop because of this inattention to a design's location on the S-curve.

Imagine that the S-curve in Figure 2 has time for its horizontal axis and velocity for its vertical axis. Given these parameters, the development of the airplane will be used to describe the six stages of design evolution.

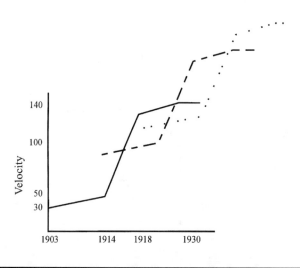

Figure 2. This S-curve can be used to describe the six stages of airplane design evolution.

Pregnancy A new concept lingers in a gestational state until the idea reaches a viable level of reality. As example, look at the many creations that tried to achieve heavier-than-air flight over the centuries.

Birth A new technological system emerges when two conditions exist:

1. There is a need for a function.
2. There are means (technology) to meet this need.

The need to establish control over the environment through mechanical systems that imitate life has long been a driving human influence, as evidenced by man's desire to fly. However, the technologies of aerodynamics and mechanics where not sufficient for the development of human flight until the late 1800s.

The technologies for the airplane were available since the development of glider flight in 1848 by Otto Lilienthal and the gasoline engine in 1859 by Etienne Lenoir. It was only because the technology of gliding alone could not overcome the safety issue when "lift" was suddenly absent (the wind dropped) that the Wright brothers devised a way of carrying an independent power source aboard their aircraft in 1903 — and a new technology got off the ground.

The need for a feature is either real/current or anticipated for the future. The initial design effort demands resources without any guarantee of an investment return. The effort is driven by faith in finding a solution which will be desired by the customer. The result is emerging technology or an application of an innovation.

Childhood   The new system appears as a result of a high-level invention. Typically, the system is primitive, inefficient, unreliable and has many unsolved problems. It does, however, provide some new function, or the means to provide the function. System development at this stage is very slow, due to lack of human and financial resources. Many design questions and issues must be answered. For example, most people may not yet be convinced of the system's usefulness. Nevertheless, a small number of enthusiasts who believe in the system's future continue to work towards its success.

The Wright brothers' first flight achieved an airspeed close to 30 mph. The airplane progressed slowly through its next developmental stages. Human and financial resources were limited because airplanes were seen as impractical curiosities. By 1913, airplane speed had increased only to 50 mph.

Adolescence (rapid development leading to independence)   This stage begins when society recognizes the value of the new system. By this time, many problems have been overcome; efficiency and performance have improved, and a new market is created. As interest in the system escalates, people and organizations invest money in development of the new product or process. This accelerates the system's development, improving the results and, in turn, attracting greater investment. A positive "feedback" loop is established, which serves to further accelerate the system's evolution.

In 1914, two stimuli for rapid growth existed. The first was the start of World War I and the military's recognition that airplanes were potentially useful. The second was increased financing and human resources as airplane design became more reliable. Airplanes were no longer expensive toys. With better financing and resources, the airplane's speed doubled from 50 to 100 mph over the four-year period from 1914 to 1918.

Maturity System development slows as the initial concept's performance reaches its natural limit. More money and labor have decreasing effects upon performance. Standard concepts, shapes and materials become established. Small improvements occur through system optimization and trade-off.

Accordingly, the rapid growth rate of the airplane has leveled off.

Decline The limits of technology have been reached and no fundamental improvement is available. The system may no longer be needed, because the function provided is no longer needed (the buggy whip). The only way to break the decline is by development of a new concept, and possibly a new technology.

By 1918, the underlying concept of the Wright airplane (biplane with ropes, textiles, low aerodynamic quality, etc.) was practically exhausted. As a result, during the next 12 to 14 years, the airplane's speed increased only to 140 mph.

The next generation (representing a new S-curve and a rebirth) began with aerodynamic, metal-framed monoplanes. This design also had a performance limit. A third S-curve started with the jet plane. New concepts with new S-curves are essential for survival in the world economy.

2. Evolution Toward Increased Ideality

Every system performs functions that generate useful and harmful effects. The general direction for system improvement toward ideality occurs by improving the ratio of useful to harmful effects.

$$Ideality = \frac{All\ Useful\ Effects}{All\ Harmful\ Effects}$$

We strive to improve the level of ideality as we create and choose inventive solutions. Chapter 5 explores the process of approaching ideality in greater detail.

An ideal design provides the required function without actually existing. A simple design using resources normally available becomes an elegant design. The Ideality equation suggests recognizing the useful and harmful effects within any design. The concept of finding the ratio has its limitations. For instance, it is difficult to quantify the cost of environmental pollution or the loss of a human life. Similarly, the ratio of increased versatility vs. usability is difficult to measure. The Analytic Hierarchy Process (AHP) helps the team rank the effects it wants to improve. Tools such as AHP ensure that the development team's analysis is both thoughtful and precise.

3. Nonuniform Development of System Elements

Each component or subsystem within a system component may have its own S-curve. Different components/subsystems usually evolve according to their own schedule. Likewise, different system components reach their inherent limits at various times. The component that reaches its limit first is "holding back" the overall system. It becomes the weak link in the design. An underdeveloped part is also a weak link. Until this component is improved the overall system's performance is limited. During the development of the airplane, psychological inertia concentrated development effort on the engine, but other aspects of the plane contained more limiting components. Understanding the interaction of all the components that influence performance is key to understanding the design. Realizing that other aspects of a system may be holding back performance is a good reason for the design team to use the Problem Formulating Process to create problems statements for these other areas.

The following case underscores the importance of focusing improvement on the weakest link in the system. A manufacturer of plastic car bumpers was producing scrap at twice the expected rate. All problem solving efforts were directed at improving the manufacturing. Since the president had been involved in formulation of the material, changes to the formula were not seriously considered. Once, out of frustration, the organization purchased a commercial formulation. The manufacturing process became stable and scrap production fell to one-tenth the targeted level. Working on the most relevant problem is key to success.

4. Evolution Toward Increased Dynamism and Controllability

The most severe beginnings of a dynamic system lack options. In the first chain-driven bicycles (one-speed coaster), the chain went from the peddle gear to the rear wheel gear. Subsequent increases in the number of gear ratios illustrate the evolutionary path from static to dynamic, from stationary to fluid or from zero degrees of freedom to infinite degrees of freedom. If you understand where the current design is, along with the customer's desire for a product that is further along this evolutionary path, you can direct development efforts intelligently. Accordingly, the bicycle became a three-speed (English style) by adjusting the internal gear ratio in the rear sprocket. The five-speed had one gear in the front and five nested gears in the rear. A cable derailer permitted shifting between rear gears. Predictably, a derailer was added to the front. More gears were added to the rear and front, establishing a common configuration of six gears in the rear and three in the front — the 18-speed bike. It should be obvious that future bicycles will shift gears automatically and have more gear ratios. The ideal design would have an infinite number of gear ratios that continuously shift to enable a single predetermined effort for any terrain. This progression can be presented as a flow chart starting from a static system, moving to a system that is changeable at a mechanical level, and ending with one that is changeable at a microlevel.

Some other structures that describe development toward increased dynamism are described below.

External Dynamization

Decrease in the degree of stability or change in an object from stationary to mobile are examples of external dynamization.

Example: Vehicles can be damaged by mail carriers, who remain in the vehicle while removing mail from postal deposit boxes (located on the street). A mailbox has been designed that is assembled on telescopic runners. The postal vehicle emits an infrared signal, which causes the mailbox to move automatically towards the vehicle to facilitate access by the carrier.

Internal Dynamization

Internal dynamization can be achieved through an increase in a system's degrees of freedom by division of the system into mobile parts.

Example: Work pieces with complex shapes can be gripped more easily by a grip "face" made up of many vertically-placed rods that are allowed horizontal movement, than by a traditional flat grip face (see Figure 7 in Chapter 4).

Example: A requirement of a bathyscaph was that it be able to descend quickly. To avoid collisions with the ocean floor, the bathyscaph's speed could be reduced by jettisoning part of the ballast. However, poor visibility made it difficult to determine the proximity of the ocean floor. An analogous problem existed with the hot air balloon. To decelerate the descent of a hot-air balloon, a heavy guide rope transferred weight from the balloon to the ground: as the balloon descended, the rope rested on the ground and thus reduced the total weight of the balloon. A guide rope in the form of a steel chain was similarly used with the bathyscaph (Figure 3).

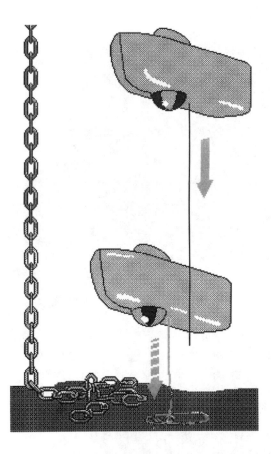

Figure 3. A chain was utilized to slow the descent of a bathyscaph as it approached the ocean floor.

Internal dynamization also can be achieved by the application of a physical effect to a system.

Example: At low temperatures, the surface area of a heat exchanger should be small to reduce heat loss. At high temperatures, the surface area should be large for quicker heat exchange.

To construct an improved heat exchanger, flaps made of titanium nickelide (a material that has shape-memory) can be attached to the exchanger. At low temperatures, the flaps are bent and lie against the body of the exchanger. At high temperatures, the flaps straighten and swing away from the exchanger, increasing its surface area (Figure 4).

Figure 4. Self-regulating heat exchanger.

Introducing a Mobile Object

Dynamization also can be increased though the introduction of a mobile object. This type of dynamization can be achieved with elements that can be interchanged, have dynamic features, or introduce adjusting elements and links.

Example: Spoilers (fins that deflect air flow) are used to improve the stability of a car at high speeds. Spoilers also increase aerodynamic drag on the car at lower speeds. Retractable back and front spoilers can provide flexible enhancement of car performance. In dry weather at speeds up to 75 mph, the front spoiler can be retracted to reduce drag. At higher speeds, it can be gradually extended to increase stability. In rainy weather at speeds more than 30 mph, both front and back spoilers can be extended to reduce skidding in water and increase stability. During braking, the back spoiler can be extended and turned to act as an air brake (Figure 5).

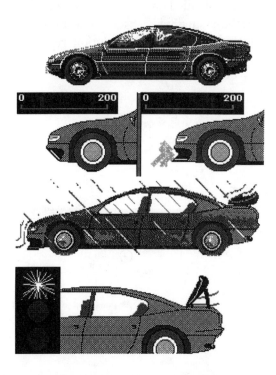

Figure 5. Retractable spoilers improve a car's performance under varying conditions.

5. Increased Complexity, then Simplification

Technological systems tend to develop first toward increased complexity (increased quantity and quality of system functions), and then toward simplification (the same or better performance is provided by a less complex system). This may be accomplished by transforming the system into a bi- or polysystem. Four possible lines of evolution for this pattern are presented.

There are many products that initially offer a new feature, and then offer several variations of the new feature or different features. For example, the first ball-point pens had one blue ink cartridge (which leaked), but subsequent designs offered writing with three different colors (variations of one feature). A stapler that incorporates a staple remover (a different feature) is a bisystem.

White-out tape offers one method of correcting documents. White-out tape in a dispenser offers variable length strips with adhesive to cover marks on standard writing paper. The dispenser can be purchased in any office supply store. Making a wider dispenser for covering large writing could be considered an extension of this latter solution. Several rolls with different widths of white tape on a single dispenser would be convenient. This is a mono- to poly-homogeneous function evolution represented by the first column in Figure 6.

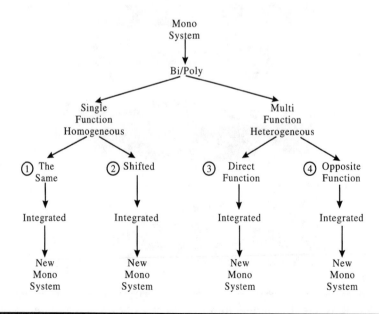

Figure 6.   Mono- to poly-homogeneous evolution.

A dispenser with different colors, textures and materials offers the same function in different formats. This is a single-function, homogeneous, shifted application of a poly-homogeneous function, represented by the second column. A dispenser which includes a stapler and knife has included different functions within the same system. This design is a multifunction, heterogeneous, direct-function system represented by the third column. Consider converting the bi- or polysystem by shifting characteristics to a system that has heterogeneous elements or inverse combinations, such as "element and anti-element" (for example, the stapler and a staple remover). Column 4 illustrates this type of evolution.

If the functions and options can be integrated, then a new monosystem has been created.

Example: The first duplication machines only duplicated. Current machines print on both sides and staple or bind the document. These machines are new monosystems. This evolution process can be used to design tomorrow's product today.

If you have a monosystem today, what are the likely bi- or polysystems of tomorrow? What will the higher level monosystem look like?

It is possible to combine patterns of evolution, as suggested earlier, by looking at combinations of patterns.

Example: Catamarans are high-stability sailboats with two hulls (a homogeneous bisystem). Binding the hulls together rigidly (static on the dynamism pattern), limits the catamaran's maneuverability. Joining the hulls with sliding joints makes the distance between hulls adjustable while creating increased maneuverability (Figure 7). This is, therefore, a dynamic homogeneous bisystem. (A secondary problem has been created for the guy wires supporting the mast.)

There is nothing to prevent you from combining the patterns of evolution to make new patterns.

Example: A nonhomogeneous bisystem with opposite functions increases hydrogenerator speed.

High speed turbine rotation is needed for efficient operation of hydrogenerators. Insufficient water flow prevents the generator rotor from reaching efficient operating speeds. If the stator could be rotated in the opposite direction, the relative speed between rotor and stator could be doubled. The water could be directed through two different channels to

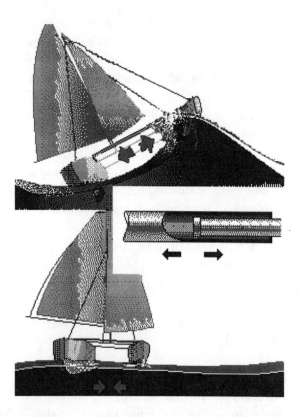

Figure 7. Joining the twin hulls of a catamaran with adjustable sliding joints makes the craft more maneuverable.

impellers for both the rotor and stator. The stator is no longer a stator, but a counter-rotor (Figure 8).

Example: A homogeneous polysystem shows that all fruit tree seeds are not equally viable. To increase the chance for a healthy, fast-growing seedling, three seeds can be planted in the same spot. Two months later, the healthiest of the three seedlings is identified and the tops are pruned away from the other two. The root stock of the pruned trees is grafted on the healthy tree. The remaining triple root system provides the seedling with more water and nutrients, ensuring fast growth (Figures 9 and 10).

Figure 8. Making the stator a counter-roter improves the operation of hydro-generators.

6. Evolution with Matching and Mismatching Components

This pattern of evolution could be called the military marching contradiction. The contradiction is resolved by using the Separation in Time principle discussed in Chapter 4. During a parade, individuals walking in unison create a powerful effect. Unfortunately, that powerful effect can destroy a bridge. Consequently, the standard operating procedure for a marching group is to allow everyone to walk with their normal gait and speed when crossing a bridge. Unison marching sets up a vibration that can destabilize a bridge.

Figures 9 and 10. A seedling can be improved by grafting its roots to the roots of two other seedlings.

Many organizations pride themselves on their understanding and application of Statistical Process Control (SPC). Such organizations take great pains to reduce variability. Their aim is to reduce variation between products, but should not necessarily be to reduce variation within products. They may be reducing the effectiveness of a product by making a subsystem symmetrical. A cutting tool with 6 cutting edges is more effective if the teeth are not exactly 60 degrees apart. A pattern of 60.5, 59, 61, 62, 58 and 59.5 would have six different frequencies thus avoiding reinforced vibration.

In this pattern of evolution, system elements are matched or mismatched to improve performance and to compensate for undesired effects. A typical evolutionary sequence can be used to explain the development of the automobile's suspension system.

   a. Unmatched elements:
      a tractor with wheels in front and tracks in back.
   b. Matched elements:
      four identical wheels on a car.
   c. Mismatched elements:
      small front and large back wheels on a dragster.
   d. Dynamic matching and mismatching:
      turning angle of left and right front wheels in expensive car.

Another simple example of mismatching components was presented in the discussion of the motor mounts (see Figure 1 in Chapter 4).

7. Evolution Toward Microlevel and Increased Use of Fields

Technological systems tend to evolve from macrosystems to microsystems. During this transition, different types of energy fields are used to achieve better performance or control. One flow in the evolution from the macrosystem to the microsystem has seven stages (Figure 11).

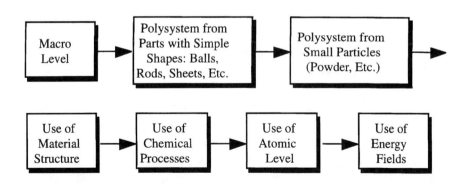

Figure 11. Evolution from a macrosystem to a microsystem in seven stages.

The evolution of a cooking oven can be described in four stages:

i. the large cast iron wood stove
ii. the smaller stove and oven fired by natural gas
iii. the electrically heated stove and oven
iv. the microwave oven.

Seven steps can be illustrated in the house construction industry:

i. macrolevel — logs
ii. simple shapes — boards
iii. small particles — particle board
iv. material structure — orientated wood chips
v. chemical — recycled plastic boards and molding
vi. atomic — dome supported by air
vii. energy fields — align ferrous particles with a magnetic field to create walls.

Cement manufacturing requires the baking of a raw material called clinker. The process uses a special horizontal, rotating kiln (a heated pipe approximately 100 meters long with a three-meter diameter). The clinker must contact a solid body to heat properly; therefore, 100 tons of steel chains are suspended inside the pipe. The chains not only effectively transfer heat, but they also grind the clinker into a wasteful fine dust. As a result, in addition to being large and expensive, a conventional cement kiln emits high levels of pollution and consumes large amounts of energy (Figure 12).

A new method for baking clinker does not produce dust. Clinker is blown from the bottom of a vertical bath filled with molten iron and the cement collects on the surface. The time during which the clinker passes through the bath is sufficient for the baking process. Cement baking furnaces using molten iron are much smaller, consume less energy and operate cleanly (Figure 13).

While function designs are going from the macro- to the microlevel, the size of a system does not necessarily decrease. As the individual sizes of systems for providing functions become smaller, more functions are being integrated. The new system with many functions may be larger than any of the individual systems. For instance, laser printers for computers have a larger footprint than earlier dot matrix printers because of additional features incorporated in the later designs.

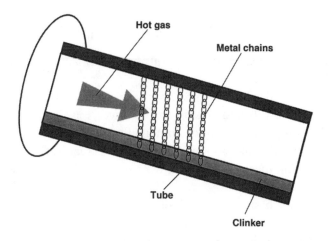

Figure 12. Using chains to heat cement.

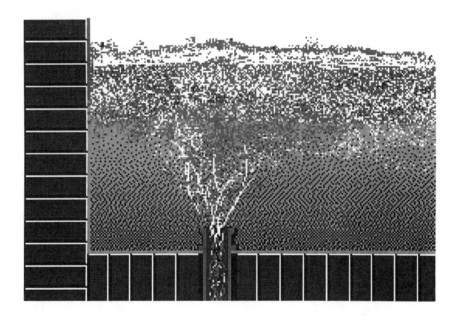

Figure 13. Cement baking furnaces using molten cast iron are much smaller, consume less energy, and operate cleanly.

8. Evolution Toward Decreased Human Involvement

Systems are developed to perform tedious functions, allowing people to do more intellectual work.

Example: A hundred years ago, clothes washing was a labor-intensive process involving a tub and washboard. Wringer washing machines were developed to reduce the amount of human effort needed to effectively clean laundry, but a considerable amount of time still was required of the operator. Automatic washing machines reduced the time and effort needed from the operator.

Once again, it is best to review all of the lines of evolution to interpolate tomorrow's designs. A clustering of alternative forecasts will show some groups with more possibilities than others. These active clusters indicate the best direction for development.

The seventh pattern of evolution referred to the increased use of fields in the future. Refer to the Chapter 6 for innovative ways of changing the field used to produce the desired function.

The last chapter offers some admonitions and guidance for a successful implementation of this revolutionary way of structured innovation called TRIZ.

> *"Human history becomes more and more a race between education and catastrophe."*
>
> — H. G. Wells
> from The Outline of History, 1920

8 Implementation

Implementation, Non-TRIZ Issues

TRIZ provides a methodology for creating solution concepts. It is not intended to develop a complete engineering solution. The solution will be within the concepts, but the details must be developed. As seen in the applications using a fast pressure drop to cause separation (Chapter 1), the concept is the same but the systems and details for each situation are different. For effective results, it is important to have clear expectations before using the TRIZ methodology.

There are many aspects to the successful implementation of the TRIZ methodology that have nothing to do with TRIZ. One of these aspects is the selection of the problem — the ISQ is very powerful when it is used in the correct problem space.

Is the product to be customer-driven or technology-driven? In either case, some aspects of a process such as Quality Function Deployment (QFD) should be used. QFD systematically translates the needs and desires of the customer and/or stakeholder into the language of the engineer. Qualities important to the customer are linked to the engineering requirements and target values of the design.

Corporate constraints must be clarified, whether they be those of the internal organization or a client. The solution may be restricted by high capital investment, production facilities or corporate image considerations.

Another area that TRIZ does not address is team composition. Many people are aware of the differences in left and right brain thinking. The left brain tends to be analytical, logical and rigid. The right brain tends to be emotional, creative and unstructured. In his book *The Creative Brain*, Ned Hermann identifies four, rather than two, behavior types.[30] Left-brain thinkers are divided into the analytic thinker dominated by the upper quadrant of the brain, and the administrator dominated by the basal quadrant. Right-brain thinkers are divided into the socially conscious, using the basal quadrant, and the creative thinker, using the upper quadrant. All four behavior types are needed during the life of a project — all four types are needed for a balanced design team. Every individual has a different dominant behavior pattern. High-level executives tend to have a good balance of all four. Engineers, on the other hand, tend to be logical, analytical and concerned little with the other three behaviors. Because of its structure, TRIZ most closely matches the thinking of the administrative mind in Hermann's model.

When project team members in high-tech organizations identify their preferred behavior type, they are almost always analytic and logical. Rarely, there will be a team member favoring one of the other quadrants. One strength of TRIZ is that it offers a structured approach to problem solving. This approach is compatible with the left-brain engineers and administrators at the same time that it helps them think creatively (right brain). The average engineer can become the better-than-average inventor/innovator.

Putting It All Together

How does it all fit together? Hundreds of TRIZ specialists have been putting it together without software for 20 years, so it must be possible. The first requirement is mastery of each tool. Practice by using the whole TRIZ tool box for each problem — even if the problem does not conveniently fall into the structure of every tool. This process will help you clarify the robustness of each tool, as well as its limitations.

An actual implementation would begin with the recording of all known information about the problem/application. Decide if you can quickly solve the problem without additional analysis. If you know the solution, then the solution is complete and probably involves only a simple Level 1 improvement.

The Innovative Situation Questionnaire (Chapter 2)

If it is a more inventive problem, completing the ISQ is the best way to start. Compiling this information often clarifies some corrective action that will resolve the problem. This solution would also be a Level 1 improvement. You again are finished.

Problem Formulation (Chapter 3)

Once the data collection is completed, the problem should be formulated using the methods described in Chapter 3. Again, clarity in the application of this process often reveals solutions to the problem without the need of knowledge-based tools. Once again, this is probably a Level 1 solution.

Large systems require 10 or 15 formulations, which can result in several hundred problem statements. Before beginning the actual formulation process, listen to the voice of the customer. What is the experience of the customer/stakeholder with the product/system? Start by opening the shipping container and doing a functional analysis of everything inside. Do the same thing with the exterior of the product — remove the exterior and look at everything on the interior. Look at subsystems. (Ask, "Why" again and again and again …)

A flow chart is made linking everything in the system to the Primary Useful Function and the Primary Harmful Function. Secondary functions are included, going to the level of detail required by the particular need that created the project. The links between nodes in the flow chart represent causal relationships. Once the chart is complete, boundaries are constructed to associate related items of subsets. These associations need not represent subsystem boundaries. A separate formulation is made for each subset.

The formulation process now begins. The output is a comprehensive set of problem statements defining the problem space. Normally, it is not possible to work with all of these problem statements, so three groupings of the problem statements are created. These groupings reflect the answers to three questions:

Group I — Which problem statements represent problems that can conveniently be solved in the immediate future?

Group II — Which problem statement represents next-generation issues?

Group III — Which problem statements are out of scope, but raise issues that may be useful in the future?

The three groupings only represent guidelines and are not to be rigidly followed. Into what grouping do you put the problem statement that generates an innovative concept leaping from the page? Identify your criteria for ranking the problem statements. These criteria must be ranked, or the implication is that all the problem statements are equally important. The Analytic Hierarchy Process (AHP) is recommended for doing this ranking. The same ranking process, possibly with different criteria, should be used for the solution concepts when they are identified.

As stated earlier, 85% of the innovative problems are solved before getting to the TRIZ methodology of analytic and knowledge-based tools. Simply using the ISQ and problem formulation process carefully can direct the team to an effective problem solution.

For those problems that have not yet revealed their solution concepts, a subset of problem statements that promise concept breakthroughs are identified.

There is overlap between several of the tools. With experience, practitioners begin to identify their favorite TRIZ tool, which becomes the one they often reach for first. For the beginner, the question is now, "What do I do next?"

Picking the Tool

Some of the tools are for analysis and some of the tools provide knowledge. Each tool has its own strengths and weaknesses because of its intended purpose.

Technical Contradiction Analysis (Chapter 4)

Type of tool — Knowledge-based.

Strength — Fast and simple to use. Offers recommendations for 1201 contradictions.

Weakness — The problem must be forced into the 39 Parameters, must be an existing system or must present a technical contradiction. Concepts are very general and limited.

Good for — ...problems defined as a contradiction that fit the format of the 39 Parameters ...problems that are technical contradictions.

Ideal Design (Chapter 5)

Type of tool — Analytic.

Strength — Establishes a vision for the future. Offers guidelines for directions to pursue. Works for new systems.

Weakness — The guidelines are general. Very dependent on experience and knowledge.

Good for — ...stimulating nontraditional thinking.

The ideal design is one component of a larger analytic tool called Algorithm for Inventive Problem Solving (ARIZ).

Su–field Analysis (Chapter 6)

Type of tool — Analytic.
Strength — Very structured. Generates very different design concepts.
Weakness — Requires a strong background in physics to identify possible fields.
Good for — …generating ideas for existing designs. …using other fields of energy and knowledge.

Patterns of Evolution (Chapter 7)

Type of tool — Knowledge-based.
Strength — Offers insights into the directions of technological evolution.
Weakness — Difficult to identify which pattern is most relevant. Difficult to find position of the current product along the evolutionary timeline.
Good for — …designing tomorrow's products today. …identifying the next generation.

This book has presented four analytic tools and two knowledge-based tools. The analytic tools give direction and the knowledge-based tools provide concepts.

There is tremendous synergy between TRIZ and two quality tools, QFD and the Taguchi Methods. By using these three approaches together, a very powerful design process results.

Synergy

The word *synergy* is derived from *synergetic*, a Greek word meaning "working together, cooperating." From start to finish, design becomes most effective when teams take advantage of the synergy between powerful tools in order to enhance their work. These three particular tools significantly improve an organization's design process: Quality Function Deployment (QFD), the Theory of Inventive Problem Solving (TRIZ) and Taguchi's methods. The first, Quality Function Deployment, translates all relevant customer information for product design into the language of the engineer. TRIZ offers a methodology for an innovative solution. Taguchi's Methods determine the parameter values for the design. Each of these tools alone has strengths and weaknesses. Together they form a complete and balanced methodology.

QFD helps an organization understand its customers and stakeholders. Founded upon precise and thorough recordings of customer responses before and after experiencing a product or service, QFD generates large amounts of information about a product's strengths and shortcomings. Since this information comes directly from the customers, it is inherently subjective. It contains the customers' demanded subjective performance criteria for a product or service, as well as descriptions of who, what, where, when and how the product is used or may be used. (See *Step by Step QFD: Customer Driven Product Design,* St. Lucie Press [CRC], 1997, for more information[31]).

The subjective ranking of the demanded performance's importance and the current judgment of satisfaction from different suppliers of the product or service provide the data to begin the design process. This information is translated into the language of the organization and is used to align the organization's resources in order to design and create the product or service the customer had only dreamed of experiencing. QFD raises a flag and indicates the precise nature of future improvements. QFD does not include any techniques intended to aid the process of innovation. In fact, to stimulate their designers' imaginations and brainstorming efforts, organizations must turn to a new tool.

The Theory of Inventive Problem Solving (TRIZ) offers quantum improvements in the creation of innovative designs. TRIZ facilitates *directed product evolution* (DPE). DPE allows an organization to design tomorrow's products today, rather than wait for this evolution to happen "naturally." However, while it is a potent tool for generating new ideas and patents, TRIZ assumes that the problem on which the design team concentrates is the appropriate one. TRIZ also does not address the issues of final design work.

Genichi Taguchi's Philosophy of Robust Design offers designers a third tool, one that helps them target particular strengths and weaknesses in a product's performance. Robust Design identifies component values which enable a design to perform on target, independently of uncontrolled influences. (See *Robust Design: Key Points for World-class Design* for more information.[32]) Three major sources of variation in performance are:

- variation in raw materials or components
- variation in the manufacturing process
- variation in the environment in which the product will be used.

For example, to direct minor flooding in a home's cellar it is now possible to glue a barrier to the floor, which is either dry or wet and which is a robust formulation.

Taguchi's method also creates product and process designs which are flexible, allowing engineers/designers to anticipate and adjust for future or variable requirements.

The combination of these three approaches yields an innovative customer-driven product design process. This complete method's superiority lies in the fact that it is not sensitive to sources of variation. Whatever uncontrolled influences may emerge, they will not negatively impact the performance of product or service.

QFD + TRIZ + Taguchi = Customer-Driven Robust Innovation

Resources

There are many easily found web sites for each of the three processes. Currently, there are two organizations providing annual symposiums which include a TRIZ component. The American Supplier Institute, located in Michigan, has a symposium in November (telephone 800-462-4500). The QFD Institute, also in Michigan, has a symposium in June (telephone 313-995-0847). Also, the bibliography at the back of this book provides a good basic reading list.

Innovation can be fun. Contradictions, no longer seen as stumbling blocks, now represent opportunities for breakthrough designs. The authors are looking forward to meeting you at a symposium or on the Internet — to share the excitement.

e-mail: john@terninko.com
web page: http://www.mv.net/ipusers/rm.

> *"Not for these I raise the song of thanks and praise;*
> *But for those obstinate questionings of sense and outward things ..."*
>
> — *William Wordsworth*
> *from "Intimations of Immortality," 1807*

Appendix A
The Application of Selected Physical Effects and Phenomena for Generating Inventive Solutions[33-34]

Required effect or property	Physical phenomenon that provides the required effect or property
1. Measuring temperature	Thermal expansion and its influence on the natural frequency of oscillations
	Thermoelectric phenomena
	Radiation spectrum
	Changes in optical, electrical and magnetic properties of substances
	Transition over the Curie point
	Hopkins, Barkhausen and Seebeck effects
2. Reducing temperature	Phase transitions
	Joule-Thomson effect
	Rank effect
	Magnetic calorie effect
	Thermoelectric phenomena
3. Increasing temperature	Electromagnetic induction
	Eddy current
	Surface effect
	Dielectric heating
	Electronic heating
	Electrical discharge
	Absorption of radiation by substances
	Thermoelectric phenomena

Required effect or property	Physical phenomenon that provides the required effect or property
4. Temperature stabilization	Phase transitions, including transition over the Curie point
5. Object location	Introduction of markers; that is, substances that are able to transform existing fields (like luminophores) or generate their own fields (like ferromagnetic materials) and therefore are easy to detect Reflection and emission of light Photo effect Deformation
	Radioactive and X-ray radiation
	Luminescence
	Changes in electric or magnetic field
	Electrical discharge
	Doppler effect
6. Moving an object	Magnetic field applied to influence an object or magnet attached to the object
	Magnetic field applied to influence a conductor with direct current passing through it
	Electric field applied to influence an electrically charged object
	Pressure transfer in a liquid or gas
	Mechanical oscillations
	Centrifugal force
	Thermal expansion
	Pressure of light
7. Moving a liquid or gas	Capillary force
	Osmosis
	Toms effect
	Waves
	Bernoulli effect
	Weissenberg effect
8. Moving an aerosol (dust particles, smoke, mist, etc.)	Electrization
	Applied electric or magnetic field
	Pressure of light
9. Formation of mixtures	Ultrasonics
	Cavitation
	Diffusion
	Applied electric field
	Magnetic field applied in combination with magnetic material
	Electrophoresis
	Solubilization
10. Separating mixtures	Electric and magnetic separation
	Electric or magnetic field applied to change the pseudoviscosity of a liquid
	Centrifugal force
	Sorption
	Diffusion
	Osmosis

Required effect or property	Physical phenomenon that provides the required effect or property
11. Stabilizating object position	Applied electric or magnetic field
	Holding a liquid by hardening through the influence of an electric or magnetic field
	Gyroscope effect
	Reactive force
12. Generating and/or manipulating force	Generating high pressure
	Applying a magnetic field through magnetic material
	Phase transitions
	Thermal expansion
	Centrifugal force
	Changing hydrostatic forces by influencing the pseudoviscosity of an electroconductive or magnetic liquid in a magnetic field
	Use of explosives
	Electrohydraulic effect
	Optical hydraulic effect
	Osmosis
13. Changing friction	Johnson-Rabeck effect
	Radiation influence
	Abnormally low friction effect
	No-wear friction effect
14. Crashing objects	Electrical discharge
	Electrohydraulic effect
	Resonance
	Ultrasonics
	Cavitation
	Use of lasers
15. Accumulating mechanical and thermal energy	Elastic deformation
	Gyroscope
	Phase transitions
16. Transferring energy through mechanical, thermal, radiation, or electric deformation	Oscillations
	Alexandrov effect
	Waves, including shock waves
	Radiation
	Thermal conductivity
	Convection
	Light reflection
	Fiber optics
	Lasers
	Electromagnetic induction
	Superconductivity
17. Influencing moving object	Applied electric or magnetic fields, with no influence through physical contact
18. Measuring dimensions	Measuring the natural frequency of oscillations
	Applying and detecting magnetic or electric markers

Required effect or property	Physical phenomenon that provides the required effect or property
19. Varying dimensions	Thermal expansion
	Deformation
	Magnetostriction
	Piezoelectric effect
20. Detecting surface properties and/or conditions	Electrical discharge
	Reflection of light
	Electronic emission
	Moiré effect
	Radiation
21. Varying surface properties	Friction
	Absorption
	Diffusion
	Bauschinger effect
	Electrical discharge
	Mechanical or acoustic oscillation
	Ultraviolet radiation.
22. Detecting volume properties and/or conditions	Introduction of marker; that is, substances that are able to transform existing fields (like luminophores) or generate their own fields (like ferromagnetic materials), depending on the properties of a material
	Changing electrical resistance, which depend on structure and/or property variations
	Interaction with light
	Electro- and/or magneto-optic phenomena
	Polarized light
	Radioactive and X-ray radiation
	Electronic paramagnetic or nuclear magnetic resonance
	Magneto-elastic effect
	Transition over the Curie point
	Hopkins and Barkhausen effect
	Ultrasonics
	Moessbauer effect
	Hall effect
23. Varying volume properties	Electric or magnetic fields applied to vary the properties of a liquid (pseudoviscosity, fluidity)
	Influencing by magnetic field via introduced magnetic material
	Heating
	Phase transitions
	Ionization by electrical field
	Ultraviolet, X-ray or radioactive radiation
	Deformation
	Diffusion
	Electric or magnetic field
	Bauschinger effect
	Thermoelectric, thermomagnetic or magneto-optic effect
	Cavitation
	Photochromatic effect
	Internal photo effect

Required effect or property	Physical phenomenon that provides the required effect or property
24. Developing certain structures, structure stabilization	Interference Standing waves Moiré effect Magnetic waves Phase transitions Mechanical and acoustic oscillation Cavitation
25. Detecting electric and magnetic fields	Osmosis Electrization Electrical discharge Piezo- and segneto-electrical effects Electrets Electronic emission Electro-optical phenomena Hopkins and Barkhausen effect Hall effect Nuclear magnetic resonance Gyromagnetic and magneto-optical phenomena
26. Detecting radiation	Optical acoustic effect Thermal expansion Photo effect Luminescence Photoplastic effect
27. Generating electromagnetic radiation	Josephson effect Induction of radiation Tunnel effect Luminescence Hann effect Cherenkov effect
28. Controlling electromagnetic fields	Use of screens Changing properties (for example, varying electrical conductivity) Changing objects shapes
29. Controlling light, light modulation	Refraction and reflection of light Electro- and magneto-optical phenomena Photo elasticity Kerr and Faraday effects Hann effect Franz-Keldysh effect
30. Initiating and intensification of chemical reactions	Ultrasonics Cavitation Ultraviolet, X-ray and radioactive radiation Electric discharge Shock waves

Appendix B
Altshuller's 39 Parameters

1. **Weight of Moving Object:** the measurable force, resulting from gravity, that a moving body exerts on the surface which prevents it from falling. A moving object is one which changes position on its own or as a result of some external force.
2. **Weight of Nonmoving Object:** the measurable force, resulting from gravity, that a stationary object exerts on the surface on which it rests. A stationary object is one which cannot change position on its own or as a result of some external force.
3. **Length of Moving Object:** the linear measure of an object's length, height or width in the direction of observed movement of that object. Movement may be caused by internal or external forces.
4. **Length of Nonmoving Object:** the linear measure of an object's length, height or width in the direction for which no observed movement occurs.
5. **Area of Moving Object:** the square measure of any plane or portion of a plane of an object which, when acted on by internal or external forces, can change its position in space.
6. **Area of Nonmoving Object:** the square measure of any plane or portion of a plane of an object which, when acted on by internal or external forces, cannot change its position in space.
7. **Volume of Moving Object:** the cubic measure of an object which can change its position in space when acted on by internal or external forces.

8. **Volume of Nonmoving Object:** the cubic measure of an object which cannot change its position in space when acted on by internal or external forces.
9. **Speed:** the rate at which an action or process is completed over time.
10. **Force:** the capacity to cause physical change in an object or system. The change may be full or partial, and permanent or temporary.
11. **Tension/Pressure:** the intensity of forces acting on an object or system, measured as the force of compression or tension per unit of area.
12. **Shape:** the outward appearance or contour of an object or system. Shape may be full or partial and permanent or temporary changes due to forces acting on the object or system.
13. **Stability of Object:** the resistance of a whole object or system to change caused by the interactions of its associated objects or systems.
14. **Strength:** under definable conditions and limits, the ability of an object or system to absorb the effects of force, speed, stress, etc. without breaking.
15. **Durability of Moving Object:** the length of time over which an object that changes position in space is able to successfully fulfill its function.
16. **Durability of Nonmoving Object:** the length of time over which an object that does not change position in space is able to successfully fulfill its function.
17. **Temperature:** the loss or addition of heat to an object or system during required functions, which may cause potentially undesirable changes to the object, system or product.
18. **Brightness:** the ratio of light energy to the area which is being lit by or in a system. Brightness includes the quality of light, degree of illumination, and other characteristics of light.
19. **Energy Spent by Moving Object:** the energy requirements of an object or system which changes position in space by its own means or by external forces.
20. **Energy Spent by Nonmoving Object:** the energy requirements of an object or system which does not change position in space in the presence of external forces.
21. **Power:** the ratio of work to the time required to perform that work. Used to measure the required but potentially undesirable changes in power evident in an object or system under given conditions.
22. **Waste of Energy:** increased inability of an object or system to exert force, especially when no work or product is produced.

23. **Waste of Substance:** decrease or elimination of material from an object or system, especially when no work or product is produced.
24. **Loss of Information:** decrease or elimination of data or input from a system.
25. **Waste of Time:** increase in the amount of time needed to complete a given action.
26. **Amount of Substance:** the number of elements or the quantity of an element used to create an object or system.
27. **Reliability:** the ability of an object or system to adequately perform its required function during some period of time or cycles.
28. **Accuracy of Measurement:** the degree to which a measurement is close to the actual value of the quantity being measured.
29. **Accuracy of Manufacturing:** the degree of correspondence between elements of an object or system to its design specifications.
30. **Harmful Factors Acting on Object:** externally produced influences acting on an object or system which reduce efficiency or quality.
31. **Harmful Side Effects:** internally produced influences acting on an object or system which reduce efficiency or quality.
32. **Ease of Manufacture:** the convenience and facility with which an object or system is produced.
33. **Ease of Use:** convenience and facility with which an object or system is operated.
34. **Ease of Repair:** convenience and facility with which an object or system is restored to operating condition after damage or extensive use.
35. **Adaptability:** the ability of an object or system to reshape or reorder itself as external conditions (environment, function, etc.) change.
36. **Complexity of Device:** the quantity and diversity of elements forming the object or system, including the relationships between elements. Complexity may also describe the difficulty of mastering the use of an object or system.
37. **Complexity of Control:** the quantity and diversity of elements used in measuring and monitoring an object or system, as well as the cost of measuring to an acceptable error.
38. **Level of Automation:** the ability of an object or system to perform operations without human interaction.
39. **Productivity:** the relationship between the number of times an operation is completed and the amount of time it takes to do it.

Appendix C
40 Principles[35]

1. **Segmentation**
   a. Divide an object into independent parts.
   b. Make an object sectional.
   c. Increase the degree of an object's segmentation.

   Example:

   1. Design sectional furniture, modular computer components or a folding wooden ruler.
   2. Garden hoses can be joined together to form any length.

2. **Extraction**
   a. Extract (remove or separate) a "disturbing" part or property from an object.
   b. Extract the only necessary part or property.

   Example:

   Using a tape recorder, reproduce a sound known to excite birds in order to scare them from the airport. (The sound is separated from the birds.)

3. **Local quality**
   a. Provide transition from a homogeneous structure of an object or outside environment (outside action) to a heterogeneous structure.
   b. Have different parts of the object carry out different functions.
   c. Place each part of the object under conditions most favorable for its operation.

Example:

1. To combat dust in coal mines, a fine mist of water in a conical form is applied to working parts of the drilling and loading machinery. The smaller the droplets, the greater an effect in combating dust, but fine mist hinders the work of the drill. The solution is to develop a layer of coarse mist around the cone of fine mist.
2. Make a pencil and an eraser in one unit.

4. Asymmetry

a. Replace a symmetrical form with an asymmetrical form of the object.
b. If an object is already asymmetrical, increase the degree of asymmetry.

Example:

1. One side of a tire is stronger than the other to withstand impact with the curb.
2. While discharging wet sand through a symmetrical funnel, the sand forms an arch above the opening causing irregular flow. A funnel of asymmetrical shape completely eliminates the arching effect.

5. Combining (Integration)

a. Combine in space homogeneous objects or objects destined for contiguous operations.
b. Combine in time homogeneous or contiguous operations.

Example:

The working element of a rotary excavator has special steam nozzles to defrost and soften the frozen ground in a single step.

6. Universality

Have the object perform multiple functions, thereby eliminating the need for some other objects.

Example:

1. Sofa converts from a sofa in the day to a bed at night.
2. Minivan seat adjusts to accommodate seating, sleeping or carrying cargo.

7. Nesting

a. Contain an object inside another, which in turn is placed inside a third object.
b. An object passes through a cavity of another object.

Example:

1. Telescoping antenna
2. Stacking chairs (on top of each other for storage)
3. Mechanical pencil with lead stored inside
4. Matrioshkas

8. Counterweight

a. Compensate for the object's weight by joining with another object that has a lifting force.
b. Compensate for the object's weight by providing aerodynamic or hydrodynamic forces.

Example:

1. Boat has hydrofoils.
2. Racing car has rear wing to increase downward pressure.

9. Prior counteraction

a. If it is necessary to carry out some action, consider a counteraction in advance.
b. If an object must be in tension, provide antitension in advance.

Example:

1. Reinforced concrete column or floor
2. Reinforced shaft — in order to make a shaft stronger, it is made out of several pipes that have been previously twisted to a calculated angle.

10. Prior action

a. Carry out the required action in advance, in full or in part.
b. Arrange objects so that they can go into action, without time loss while waiting for the action (and from the most convenient position).

Example:

1. Utility knife blade made with a groove allowing the dull part of the blade to be broken off to restore sharpness.
2. Rubber cement in a bottle is difficult to apply neatly and uniformly. Instead, it is formed into a tape so that the proper amount can be applied more easily.

11. Cushion in advance

Compensate for the relatively low reliability of an object by countermeasures taken in advance.

Example:

To prevent shoplifting, the owner of a store attaches a special tag containing a magnetized plate. In order for the customer to carry the merchandise out of the store, the plate is demagnetized by the cashier.

12. Equipotentiality

Change the condition of work so that an object need not be raised or lowered.

Example:

Automobile engine oil is changed by pit crew (so that expensive lifting equipment is not needed).

13. Inversion

a. Instead of an action dictated by the specifications of the problem, implement an opposite action.
b. Make object a moving part, or make nonmoving part movable and outside environment immovable.
c. Turn the object upside down.

Example:

Abrasively clean parts by vibrating the parts instead of the abrasive.

14. Spheroidality

a. Replace linear parts or flat surfaces with curved ones, and cubical shapes with spherical shapes.
b. Use rollers, balls and spirals.

c. Replace a linear motion with a rotating motion, utilize a centrifugal force.

Example:

A computer mouse utilizes ball construction to transfer linear, two-axis motion into a vector.

15. Dynamicity

a. Make characteristics of an object or outside environment automatically adjust for optimal performance at each stage of operation.
b. Divide an object into elements able to change position relative to each other.
c. If an object is immovable, make it movable or interchangeable.

Example:

1. A flashlight can have a flexible gooseneck between the body and the lamp head.
2. A transport vessel has a body of cylindrical shape. To reduce the draft of the vessel under full load, the body is comprised of two hinged half-cylindrical parts that can be opened.

16. Partial, overdone, or excessive action

If it is difficult to obtain 100% of a desired effect, achieve somewhat more or less to greatly simplify the problem.

Example:

1. A cylinder is painted by dipping it into paint, but it is covered by more paint than is desired. Excess paint is removed by rapidly rotating the cylinder.
2. To obtain uniform discharge of a metallic powder from a bin, the hopper has a special internal funnel which is continually overfilled to provide nearly constant pressure.

17. Moving to a new dimension

a. Remove problems of moving an object in a line by allowing two-dimensional movement (along a plane). Similarly, problems in moving an object in a plane are removed if the object can be changed to allow three-dimensional movement.
b. Use a multilayer assembly of objects instead of a single layer.

c. Incline the object or turn it "on its side."
d. Project images on to neighboring areas or on to the reverse side of the object.

Example:

A greenhouse has a concave reflector on the northern part of the house to improve illumination during the day by reflecting sunlight into that part of the house.

18. Mechanical vibration

a. Set an object into oscillation.
b. If oscillation exists, increase its frequency, even as far as ultrasonic.
c. Use the frequency of resonance.
d. Instead of mechanical vibrators, use piezovibrators.
e. Use ultrasonic vibrations in conjunction with an electromagnetic field.

Example:

1. To remove a cast from the body without skin injury, a conventional hand saw is replaced with a vibrating knife.
2. Vibrate a casting mold while it is being filled to improve flow and structural properties.

19. Periodic action

a. Replace a continuous action with a periodic one (impulse).
b. If an action is already periodic, change its frequency.
c. Use pauses between impulses to provide additional action.

Example:

1. An impact wrench loosens corroded nuts using impulses rather than a continuous force.
2. A warning lamp flashes so that it is even more noticeable than if continuously lit.

20. Continuity of useful action

a. Carry out an action without a break — all parts of an object should be constantly operating at full capacity.
b. Remove an idle and intermediate motion.

Example:

> A drill can have cutting edges which allow for the cutting process in forward and reverse directions.

21. Rushing through

Perform harmful or hazardous operations at very high speed.

Example:

> A cutter for thin-wall plastic tubes prevents tube deformation during cutting by running at a very high speed (cuts before the tube has a chance to deform).

22. Convert harm into benefit

a. Utilize a harmful factor or harmful effect of an environment to obtain a positive effect.
b. Remove a harmful factor by combining it with another harmful factor.
c. Increase the amount of harmful action until it ceases to be harmful.

Example:

> 1. Sand or gravel freezes solid when transported through cold climates. Over freezing (using liquid nitrogen) embrittles the ice, which permits pouring.
> 2. When using high-frequency current to heat metal, only the outer layer is heated. This negative effect is now used for surface heat treating.

23. Feedback

a. Introduce feedback.
b. If feedback already exists, reverse it.

Example:

> 1. Water pressure from a well is maintained by sensing output pressure and turning on a pump if pressure is low.
> 2. Ice and water are measured separately but must be combined to an exact total weight. Because it is difficult to precisely dispense the ice, it is measured first. The weight of the ice is fed to the water control, which precisely dispenses the amount of water needed.

3. Noise canceling devices sample noise signals, phase shift them and feed them back to cancel the effect of the noise source.

24. Mediator

  a. Use an intermediary object to transfer or carry out an action.
  b. Temporarily connect an object to another one that is easy to remove.

  Example:

  > To reduce energy loss when applying current to liquid metal, use cooled electrodes and intermediate liquid metals with a lower melting temperature.

25. Self-service

  a. Make the object service itself and carry out supplementary and repair operations.
  b. Make use of waste material and energy.

  Example:

  > 1. To distribute an abrasive material evenly on the face of crushing rollers and to prevent feeder wear, its surface is made out of the same abrasive material.
  > 2. In an electric welding gun, the rod is advanced by a special device. To simplify the system, the rod is advanced by a solenoid controlled by the welding current.

26. Copying

  a. Use a simple or inexpensive copy instead of an object which is complex, expensive, fragile or inconvenient to use.
  b. Replace an object or system of objects by an optical copy or image. A scale can be used to reduce or enlarge the image.
  c. If visible optical copies are used, replace them with infrared or ultraviolet copies.

  Example:

  > The height of tall objects can be determined by measuring their shadows.

27. An inexpensive short-life object instead of an expensive durable one

  Replace an expensive object by a collection of inexpensive ones, compromising other properties (longevity, for instance).

Example:

1. Diapers that are disposable.
2. A disposable mousetrap consists of a baited plastic tube. A mouse enters the trap through a cone-shaped opening. The angled walls do not allow the mouse to exit.

28. Replacement of a mechanical system

a. Replace a mechanical system by an optical, acoustical or odor system.
b. Use an electrical, magnetic or electromagnetic field for interaction with the object.
c. Replace fields.
d. Use a field in conjunction with ferromagnetic particles.

Example:

1. Stationary fields change to moving fields.
2. Fixed fields become fields that change in time.
3. Random fields change to structured ones.
4. To increase a bond of metal coating to a thermoplastic material, the process is carried out inside an electromagnetic field to apply force to the metal.

29. Use a pneumatic or hydraulic construction

Replace solid parts of an object with gas or liquid. These parts can use air or water for inflation or use air or hydrostatic cushions.

Example:

1. To increase the draft of an industrial chimney, a spiral pipe with nozzles is installed. When air flows through the nozzles, it creates an airlike wall which reduces drag.
2. For shipping fragile products, air-bubble envelopes or foamlike materials are used.

30. Flexible film or thin membranes

a. Replace customary construction with flexible membrane and thin film.
b. Isolate an object from the outside environment with a thin film or fine membrane.

Example:

> To prevent the loss of water evaporating from the leaves of plants, polyethylene spray is applied. The polyethylene hardens and plant growth improves because the polyethylene film passes oxygen better than water vapor.

31. Use of porous material

   a. Make an object porous or use additional porous elements (inserts, covers, etc.).
   b. If an object is already porous, fill the pores in advance with some substance.

   Example:

   > To avoid pumping coolant to a machine, some parts of the machine are filled with porous material (porous powdered steel) soaked in coolant liquid which evaporates when the machine is working, providing short-term, uniform cooling.

32. Changing the color

   a. Change the color of an object or its surroundings.
   b. Change the translucency of an object or its surroundings.
   c. Use color additives to observe difficult to see objects or processes.
   d. If such additives are already used, employ luminescent traces or tracer elements.

   Example:

   1. A transparent bandage enables a wound to be inspected without the dressing being removed.
   2. In steel mills, a water curtain is used to protect workers from overheating. But this curtain only protects from infrared rays, so the bright light from the melted steel can easily get through the curtain. A coloring is added to the water to create a filter effect while remaining transparent.

33. Homogeneity

   Make objects interact with a primary object of the same material, or a material similar in behavior.

Example:

> The surface of a feeder for abrasive grain is made of the same material that runs through the feeder, allowing continuous restoration of the surface without it being worn out.

34. Rejecting and regenerating parts

  a. After it has completed its function or become useless, reject or modify (e.g., discard, dissolve or evaporate) an element of an object.
  b. Restore directly any used-up part of an object.

Example:

> 1. Bullet casings are ejected after the gun fires.
> 2. Rocket boosters separate after serving their function.

35. Transformation of physical and chemical states of an object

Change the aggregate state of an object, the concentration of density, the degree of flexibility or the temperature.

Example:

> In a system for brittle friable materials, the surface of the spiral feedscrew is made from an elastic material with two spiral springs. In order to control the process, the pitch of the screw can be changed remotely.

36. Phase transition

Implement an effect developed during the phase transition of a substance. For instance, during the change of volume or during liberation or absorption of heat.

Example:

> To control the expansion of ribbed pipes, they are filled with water and cooled to a freezing temperature.

37. Thermal expansion

  a. Use expansion or contraction of a material by heat.
  b. Use various materials with different coefficients of heat expansion.

Example:

To control the opening of roof windows in a greenhouse, bi-metallic plates are connected to the windows. With a change of temperature, the plates bend and make the window open or close.

38. Use strong oxidizers

a. Replace normal air with enriched air.
b. Replace enriched air with oxygen.
c. In oxygen or in air, treat a material with ionizing radiation.
d. Use ionized oxygen.

Example:

To obtain more heat from a torch, oxygen is fed to the torch instead of atmospheric air.

39. Inert environment

a. Replace the normal environment with an inert one.
b. Carry out a process in a vacuum.

Example:

To prevent cotton from catching fire in a warehouse, it is treated with inert gas during transport to the storage area.

40. Composite materials

Replace a homogeneous material with a composite one.

Example:

Military aircraft wings are made of composites of plastics and carbon fibers for high strength and low weight.

Appendix D
40 Principles in Order According to Frequency of Use[36]

35. Transformation of physical and chemical states of an object
10. Prior action
1. Segmentation
28. Replacement of a mechanical system
2. Extraction
15. Dynamicity
19. Periodic action
18. Mechanical vibration
32. Changing the color
13. Inversion
26. Copying
3. Local quality
27. An inexpensive short-life object instead of an expensive durable one
29. Use a pneumatic or hydraulic construction
34. Rejecting and regenerating parts
16. Partial or overdone action
40. Composite materials
24. Mediator
17. Moving to a new direction
6. Universality
14. Spheroidality
22. Convert harm to benefit
39. Inert environment

4. Asymmetry
30. Flexibility
37. Thermal expansion
36. Phase transition
25. Self-service
11. Cushion in advance
31. Use of porous materials
38. Use strong oxidizers
8. Counterweight
5. Combining
7. Nesting
21. Rushing through
23. Feedback
12. Equipotentiality
33. Homogeneity
9. Prior counteraction
20. Continuity of useful action

Appendix E
The Contradiction Table[37, 38]

Feature to Change \ Undesired Result (Conflict)	1 Weight of moving object	2 Weight of non-moving object	3 Length of moving object	4 Length of non-moving object	5 Area of moving object	6 Area of non-moving object	7 Volume of moving object	8 Volume of non-moving object	9 Speed	10 Force	11 Tension, pressure	12 Shape	13 Stability of object
1 Weight of moving object			15,8, 29,34		29,17, 38,34		29,2, 40,28		2,8, 15,38	8,10, 18,37	10,36, 37,40	10,14, 35,40	1,35, 19,39
2 Weight of non-moving object				10,1, 29,35		35,30, 13,2		5,35, 14,2		8,10, 19,35	13,29, 10,18	13,10, 29,14	26,39, 1,40
3 Length of moving object	8,15, 29,34				15,17, 4		7,17, 4,35		13,4, 8	17,10, 4	1,8, 35	1,8, 10,29	1,8, 15,34
4 Length of non-moving object		35,28, 40,29				17,7, 10,40		35,8, 2,14		28,10	1,14, 35	13,14, 15,7	39,37, 35
5 Area of moving object	2,17, 29,4		14,15, 18,4				7,14, 17,4		29,30, 4,34	19,30, 35,2	10,15, 36,28	5,34, 29,4	11,2, 13,39
6 Area of non-moving object		30,2, 14,18		26,7, 9,39						1,18, 35,36	10,15, 36,37		2,38
7 Volume of moving object	2,26, 29,40		1,7, 4,35		1,7, 4,17				29,4, 38,34	15,35, 36,37	6,35, 36,37	1,15, 29,4	28,10, 1,39
8 Volume of non-moving object		35,10, 19,14	19,14	35,8, 2,14						2,18, 37	24,35	7,2, 35	34,28, 35,40
9 Speed	2,28, 13,38		13,14, 8		29,30, 34		7,29, 34			13,28, 15,19	6,18, 38,40	35,15, 18,34	28,33, 1,18
10 Force	8,1, 37,18	18,13, 1,28	17,19, 9,36	28,10	19,10, 15	1,18, 36,37	15,9, 12,37	2,36, 18,37	13,28, 15,12		18,21, 11	10,35, 40,34	35,10, 21
11 Tension, pressure	10,36, 37,40	13,29, 10,18	35,10, 36	35,1, 14,16	10,15, 36,25	10,15, 35,37	6,35, 10	35,24	6,35, 36	36,35, 21		35,4, 15,10	35,33, 2,40
12 Shape	8,10, 29,40	15,10, 26,3	29,34, 5,4	13,14, 10,7	5,34, 4,10		14,4, 15,22	7,2, 35	35,15, 34,18	35,10, 37,40	34,15, 10,14		33,1, 18,4
13 Stability of object	21,35, 2,39	26,39, 1,40	13,15, 1,28		37	2,11, 13	39	28,10, 19,39	34,28, 35,40	33,15, 28,18	10,35, 21,16	2,35, 40	22,1, 18,4
14 Strength	1,8, 40,15	40,26, 27,1	1,15, 8,35	15,14, 28,26	3,34, 40,29	9,40, 28	10,15, 14,7	9,14, 17,15	8,13, 26,14	10,18, 3,14	10,3, 18,40	10,30, 35,40	13,17, 35
15 Durability of moving object	19,5, 34,31		2, 19, 9		3,17, 19		10,2, 19,30		3, 35, 5	19,2, 16	19,3, 27	14,26, 28,25	13,3, 35
16 Durability of non-moving object		6,27, 19,16		1,10, 35				35,34, 38					39,3, 35,23
17 Temperature	36,22, 6,38	22,35, 32	15,19, 9	15,19, 9	3,35, 39,18	35,38	34,39, 40,18	35,6, 4	2,28, 36,30	35,10, 3,21	35,39, 19,2	14,22, 19,32	1,35, 32
18 Brightness	19,1, 32	2,35, 32	19,32, 16		19,32, 26		2,13, 10		10,13, 19	26,19, 6		32,30	32,3, 27
19 Energy spent by moving object	12,18, 28,31		12,28		15,19, 25		35,13, 18		8,15, 35	16,26, 21,2	23,14, 25	12,2, 29	19,13, 17,24
20 Energy spent by non-moving object		19,9, 6,27								36,37			27,4, 29,18

Appendix E

	Feature to Change \ Undesired Result (Conflict)	14 Strength	15 Durability of moving object	16 Durability of non-moving object	17 Temperature	18 Brightness	19 Energy spent by moving object	20 Energy spent by non-moving object	21 Power	22 Waste of energy	23 Waste of substance	24 Loss of information	25 Waste of time	26 Amount of substance
1	Weight of moving object	28,27, 18,40	5,34, 31,35		6,20, 4,38	19,1, 32	35,12, 34,31		12,36, 18,31	6, 2, 34,19	5,35, 3,31	10,24, 35	10,35, 20,28	3,26, 18,31
2	Weight of non-moving object	28,2, 10,27			2,27, 19,6	28,19, 32,22	19,32, 35		18,19, 28,1	15,19, 18,22	18,19, 28,15	5, 8, 13,30	10,15, 35	10,20, 35,26	19,6, 18,26
3	Length of moving object	8,35, 29,34	19		10,15, 19	32	8,35, 24		1,35	7, 2, 35,39	4,29, 23,10	1, 24	15, 2, 29	29, 35
4	Length of non-moving object	15,14, 28,26		1,40, 35	3,35, 38,18	3,25			12,8	6,28	10,28, 24,35	24,26	30,29, 14	
5	Area of moving object	3,15, 40,14	6,3		2,15, 16	15,32, 19,13	19,32		19,10, 32,18	15,17, 30,26	10,35, 2,39	30,26	26, 4	29,30, 6,13
6	Area of non-moving object		40		2,10, 19,30	35,39, 38			17,32	17,7, 30	10,14, 18,39	30,16	10,35, 4,18	2, 18, 40,4
7	Volume of moving object	9,14, 15,7	6,35, 4		34,39, 10,18	2,13, 10	35		35,6, 13,18	7,15, 13,16	36,39, 34,10	2, 22	2, 6, 34,10	29,30, 7
8	Volume of non-moving object	9,14, 17,15		35,34, 38	35, 6, 4				30,6		10,39, 35,34		35,16, 32,18	35, 3
9	Speed	8,3, 26,14	3,19, 35,5		28,30, 36,2	10,13, 19	8,15, 35,38		19,35, 38,2	14,20, 19,35	10,13, 28,38	13, 26		18,19, 29,38
10	Force	35,10, 14,27	19,2		35,10, 21		19,17, 10	1,16, 36,37	19,35, 18,37	14,15	8,35, 40,5		10,37, 36	14,29, 18,36
11	Tension, pressure	9,18, 3,40	19,3, 27		35,39, 19,2		14,24, 10,37		10,35, 14	2,36, 25	10,36, 3,37		37,36, 4	10,14, 36
12	Shape	30,14, 10,40	14,26, 9,25		22,14, 19,32	13,15, 32	2,6, 34,14		4, 6, 2	14	35,29, 3, 5		14,10, 34,17	36, 22
13	Stability of object	17,9, 15	13,27, 10,35	39,3, 35,23	35,1, 32	32,3, 27,15	13,19	27,4, 29,18	32,35, 27,31	14,2, 39,6	2, 14, 30,40		35,27	15,32, 35
14	Strength		27,3, 26		30,10, 40	35,19	19,35, 10	35	10,26, 35,28	35	35,28, 31,40		29,3, 28,10	29,10, 27
15	Durability of moving object	27,3, 10			19,35, 39	2,19, 4,35	28,6, 35,18		19,10, 35,38		28,27, 3,18	10	20,10, 28,18	3, 35, 10,40
16	Durability of non-moving object				19,18, 36,40				16		27,16, 18,38	10	28,20, 10,16	3, 35, 31
17	Temperature	10,30, 22,40	19,13, 39	19,18, 36,40			32,30, 21,16	19,15, 3,17	2,14, 17,25	21,17, 35,38	21,36, 29,31		35,28, 21,18	3, 17, 30,39
18	Brightness	35,19	2, 19, 6		32,35, 19		32,1, 19	32,35, 1,15	32	19,16, 1, 6	13, 1	1, 6	19, 1, 26,17	1, 19
19	Energy spent by moving object	5,19, 9,35	28,35, 6,18		19,24, 3,14	2,15, 19			6,19, 37,18	12,22, 15,24	35,24, 18,5		35,38, 19,18	34,23, 16,18
20	Energy spent by non-moving object	35					19,2, 35,32				28,27, 18,31			3, 35, 31

	Undesired Result (Conflict) / Feature to Change	27 Reliability	28 Accuracy of measurement	29 Accuracy of manufacturing	30 Harmful factors acting on object	31 Harmful side effects	32 Manufacturability	33 Convenience of use	34 Repairability	35 Adaptability	36 Complexity of device	37 Complexity of control	38 Level of automation	39 Productivity
1	Weight of moving object	3, 11, 1, 27	28,27, 35,26	28,35, 26,18	22,21, 18,27	22,35, 31,39	27,28, 1,36	35,3, 2,24	2,27, 28,11	29,5, 15,8	26,30, 36,34	28,29, 26,32	26,35, 18,19	35,3, 24,37
2	Weight of non-moving object	10,28, 8, 3	18,26, 28	10,1, 35,17	2, 19, 22,37	35,22, 1,39	28, 1, 9	6,13, 1, 32	2,27, 28,11	19,15, 29	1,10, 26,39	25,28, 17,15	2, 26, 35	1, 28, 15,35
3	Length of moving object	10,14, 29,40	28,32, 4	10,28, 29,37	1,15, 17,24	17,15	1, 29, 17	15,29, 35,4	1, 28, 10	14,15, 1,16	1, 19, 26,24	35,1, 26,24	17,24, 26,16	14,4, 28,29
4	Length of non-moving object	15,29, 28	32,28, 3	2, 32, 10	1, 18		15, 17, 27	2, 25	3	1, 35	1, 26	26		30,14, 7,26
5	Area of moving object	29, 9	26,28, 32,3	2,32	22,33, 28,1	17,2, 18,39	13,1, 26,24	15,17, 13,16	15,13, 10,1	15, 30	14, 1, 13	2,36, 26,18	14,30, 28,23	10,26, 34,2
6	Area of non-moving object	32,35, 40,4	26,28, 32,3	2,29, 18,36	27,2, 39,35	22, 1, 40	40, 16	16, 4	16	15, 16	1, 18, 36	2,35, 30,18	23	10,15, 17,7
7	Volume of moving object	14, 1, 40,11	25,26, 28	25,28, 2,16	22,21, 27,35	17,2, 40,1	29, 1, 40	15,13, 30,12	10	15, 29	26, 1	29,26, 4	35,34, 16,24	10, 6, 2,34
8	Volume of non-moving object	2,35, 16		35,10, 25	34,39, 19,27	30,18, 35,4	35		1		1, 31	2, 17, 26		35,37, 10,2
9	Speed	11,35, 27,28	28,32, 1,24	10,28, 32,25	1,28, 35,23	2,24, 35,21	35,13, 8,1	32,28, 13,12	34,2, 28,27	15,10, 26	10,28, 4,34	3,34, 27,16	10, 18	
10	Force	3,35, 13,21	35,10, 23,24	28,29, 37,36	1,35, 40,18	13,3, 36,24	15,37, 18,1	1,28, 3,25	15, 1, 11	15,17, 18,20	26,35, 10,18	36,37, 10,19	2, 35	3,28, 35,37
11	Tension, pressure	10,13, 19,35	6, 28, 25	3, 35	22, 2, 37	2, 33, 27, 18	1, 35, 16	11	2	35	19, 1, 35	2, 36, 37	35, 24	10,14, 35,37
12	Shape	10,40, 16	28,32, 1	32,30, 40	22,1, 2, 35	35, 1	1,32, 17,28	32,15, 26	2, 13, 1	1, 15, 29	16,29, 1,28	15,13, 39	15, 1, 32	17,26, 34,10
13	Stability of object		13	18	35,24, 30,18	35,40, 27,39	35, 19	32,35, 30	2,35, 10,16	35,30, 34,2	2,35, 22,26	35,22, 39,23	1, 8, 35	23,35, 40,3
14	Strength	11, 3	3, 27, 16	3, 27	18,35, 37,1	15,35, 22,2	11,3, 10,32	32,40, 28,2	27,11, 3	15, 3, 32	2, 13, 28	27, 3, 15, 40	15	29,35, 10,14
15	Durability of moving object	11, 2, 13	3	3,27, 16,40	22,15, 33,28	21,39, 16,22	27, 1, 4	12, 27	29,10, 27	1, 35, 13	10,4, 29,15	19,29, 39,35	6, 10	35,17, 14,19
16	Durability of non-moving object	34,27, 6,40	10, 26, 24		17,1, 40,33	22	35, 10	1	1	2		25,34, 6,35	1	10,20, 16,38
17	Temperature	19,35, 3,10	32,19, 24	24	22,33, 35,2	22,35, 2,24	26, 27	26, 27	4, 10, 16	2,18, 27	2,17, 16	3,27, 35,31	26,2, 19,16	15,28, 35
18	Brightness		11,15, 32	3, 32	15, 19	35,19, 32,39	19,35, 28,26	28,26, 19	15,17, 13,16	15, 1, 1, 19	6, 32, 13	32, 15	2, 26, 10	2, 25, 16
19	Energy spent by moving object	19,21, 11,27	3, 1, 32		1,35, 6,27	2, 35, 6	28,26, 30	19, 35	1,15, 17,28	15,17, 13,16	2, 29, 27,28	35, 38	32, 2	12,28, 35
20	Energy spent by non-moving object	10,36, 23			10, 2, 22,37	19,22, 18	1, 4					19,35, 16,25		1, 6

Appendix E

	Feature to Change \ Undesired Result (Conflict)	1 Weight of moving object	2 Weight of non-moving object	3 Length of moving object	4 Length of non-moving object	5 Area of moving object	6 Area of non-moving object	7 Volume of moving object	8 Volume of non-moving object	9 Speed	10 Force	11 Tension, pressure	12 Shape	13 Stability of object
21	Power	8,36, 38,31	19,26, 17,27	1,10, 35,37		19,38	17,32, 13,38	35,6, 38	30,6, 25	15,35, 2	26,2, 36,35	22,10, 35	29,14, 2,40	35,32, 15,31
22	Waste of energy	15,6, 19,28	19,6, 18,9	7,2, 6,13	6,38, 7	15,26, 17,30	17,7, 30,18	7,18, 23	7	16,35, 38	36,38			14,2, 39,6
23	Waste of substance	35,6, 23,40	35,6, 22,32	14,29, 10,39	10,28, 24	35,2, 10,31	10,18, 39,31	1,29, 30,36	3,39, 18,31	10,13, 28,38	14,15, 18,40	3,36, 37,10	29,35, 3,5	2,14, 30,40
24	Loss of information	10,24, 35	10,35, 5	1,26	26	30,26	30,16		2,22	26,32				
25	Waste of time	10,20, 37,35	10,20, 26,5	15,2, 29	30,24, 14,5	26,4, 5,16	10,35, 17,4	2,5, 34,10	35,16, 32,18		10,37, 36,5	37,36, 4	4,10, 34,17	35,3, 22,5
26	Amount of substance	35,6, 18,31	27,26, 18,35	29,14, 35,18		15,14, 29	2,18, 40,4	15,20, 29		35,29, 34,28	35,14, 3	10,36, 14,3	35,14	15,2, 17,40
27	Reliability	3,8, 10,40	3,10, 8,28	15,9, 14,4	15,29, 28,11	17,10, 14,16	32,35, 40,4	3,10, 14,24	2,35, 24	21,35, 11,28	8,28, 10,3	10,24, 35,19	35,1, 16,11	
28	Accuracy of measurement	32,35, 26,28	28,35, 25,26	28,26, 5,16	32,28, 3,16	26,28, 32,3	26,28, 32,3	32,13, 6		28,13, 32,24	32,2	6,28, 32	6,28, 32	32,35, 13
29	Accuracy of manufacturing	28,32, 13,18	28,35, 27,9	10,28, 29,37	2,32, 10	28,33, 29,32	2,29, 18,36	32,28, 2	25,10, 35	10,28, 32	28,19, 34,36	3,35	32,30, 40	30,18
30	Harmful factors acting on object	22,21, 27,39	2,22, 13,24	17,1, 39,4	1,18	22,1, 33,28	27,2, 39,35	22,23, 37,35	34,39, 19,27	21,22, 35,28	13,35, 39,18	22,2, 37	22,1, 3,35	35,24, 30,18
31	Harmful side effects	19,22, 15,39	35,22, 1,39	17,15, 16,22		17,2, 18,39	22,1, 40	17,2, 40	30,18, 35,4	35,28, 3,23	35,28, 1,40	2,33, 27,18	35,1	35,40, 27,39
32	Manufacturability	28,29, 15,16	1,27, 36,13	1,29, 13,17	15,17, 27	13,1, 26,12	16,40	13,29, 1,40	35	35,13, 8,1	35,12	35,19, 1,37	1,28, 13,27	11,13, 1
33	Convenience of use	25,2, 13,15	6,13, 1,25	1,17, 13,12		1,17, 13,16	18,16, 15,39	1,16, 35,15	4,18, 39,31	18,13, 34	28,13, 35	2,32, 12	15,34, 29,28	32,35, 30
34	Repairability	2,27, 35,11	2,27, 35,11	1,28, 10,25	3,18, 31	15,13, 32	16,25	25,2, 35,11	1	34,9	1,11, 10	13	1,13, 2,4	2,35
35	Adaptability	1,6, 15,8	19,15, 29,16	35,1, 29,2	1,35, 16	35,30, 29,7	15,16	15,35, 29		35,10, 14	15,17, 20	35,16	15,37, 1,8	35,30, 14
36	Complexity of device	26,30, 34,36	2,36, 35,39	1,19, 26,24	26	14,1, 13,16	6,36	34,25, 6	1,16	34,10, 28	26,16	19,1, 35	29,13, 28,15	2,22, 17,19
37	Complexity of control	27,26, 28,13	6,13, 28,1	16,17, 26,24	26	2,13, 15,17	2,39, 30,16	29,1, 4,16	2,18, 26,31	3,4, 16,35	36,28, 40,19	35,36, 37,32	27,13, 1,39	11,22, 39,30
38	Level of automation	28,26, 18,35	28,26, 35,10	14,13, 17,28	23	17,14, 13		35,13, 16		28,10	2,35	13,35	15,32, 1,13	18,1
39	Productivity	35,26, 24,37	28,27, 15,3	18,4, 28,38	30,7, 14,26	10,26, 34,31	10,35, 17,7	2,6, 34,10	35,37, 10,2		28,15, 10,36	10,37, 14	14,10, 34,40	35,3, 22,39

Feature to Change \ Undesired Result (Conflict)	14 Strength	15 Durability of moving object	16 Durability of non-moving object	17 Temperature	18 Brightness	19 Energy spent by moving object	20 Energy spent by non-moving object	21 Power	22 Waste of energy	23 Waste of substance	24 Loss of information	25 Waste of time	26 Amount of substance
21 Power	26,10, 28	19,35, 10,38	16	2,14, 17,25	16,6, 19	16,6, 19,37			10,35, 38	28,27, 18,38	10, 19	35,20, 10,6	4, 34, 19
22 Waste of energy	26			19,38, 7	1,13, 32,15			3,38		35,27, 2,37	19, 10	10,18, 32,7	7, 18, 25
23 Waste of substance	35,28, 31,40	28,27, 3,18	27,16, 18,38	21,36, 39,31	1, 6, 13	35,18, 24,5	28,27, 12,31	28,27, 18,38	35,27, 2,31			15,18, 35,10	6, 3, 10,24
24 Loss of information		10	10		19			10,19	19,10			24,26, 28,32	24,28, 35
25 Waste of time	29,3, 28,18	20,10, 28,18	28,20, 10,16	35,29, 21,18	1,19, 26,17	35,38, 19,18	1	35,20, 10,6	10,5, 18,32	35,18, 10,39	24,26, 28,32		35,38, 18,16
26 Amount of substance	14,35, 34,10	3,35, 10,40	3,35, 31	3,17, 39		34,29, 16,18	3,35, 31	35	7,18, 25	6,3, 10,24	24,28, 35	35,38, 18,16	
27 Reliability	11,28	2,35, 3,25	34,27, 6,40	3,35, 10	11,32, 13	21,11, 27,19	36,23	21,11, 26,31	10,11, 35	10,35, 29, 39	10,28	10,30, 4	21,28, 40,3
28 Accuracy of measurement	28,6, 32	28,6, 32	10,26, 24	6,19, 28,24	6, 1, 32	3, 6, 32		3, 6, 32	26,32, 27	10,16, 31,28		24,34, 28,32	2, 6, 32
29 Accuracy of manufacturing	3, 27	3,27, 40		19,26	3,32	32,2		32,2	13,32, 2	35,31, 10,24		32,26, 28,18	32,30
30 Harmful factors acting on object	18,35, 37,1	22,15, 33,28	17,1, 40,33	22,33, 35,2	1,19, 32,13	1,24, 6,27	10,2, 22,37	19,22, 31,2	21,22, 35,2	33,22, 19,40	22,10, 2	35,18, 34	35,33, 29,31
31 Harmful side effects	15,35, 22,2	15,22, 33,31	21,39, 16,22	22,35, 2,24	19,24, 39,32	2,35, 6	19,22, 18	2,35, 18	21,35, 2,22	10,1, 34	10,21, 29	1,22	3,24, 39,1
32 Manufacturability	1,3, 10,32	27, 1, 4	35,16	27,26, 18	28,24, 27,1	28,26, 27,1	1,4	27,1, 12,24	19,35	15,34, 33	32,24, 18,16	35,28, 34,4	35,23, 1,24
33 Convenience of use	32,40, 3,28	29,3, 8,25	1,16, 25	26,27, 13	13,17, 1,24	1,13, 24		35,34, 2,10	2,19, 13	28,32, 2,24	4,10, 27,22	4,28, 10,34	12,35
34 Repairability	11,1, 2,9	11,29, 28,27	1	4,10	15,1, 13	15,1, 28,16		15,10, 32,2	15,1, 32,19	2,35, 34,27		32,1, 10,25	2,28, 10,25
35 Adaptability	35,3, 32,6	13,1, 35	2,16	27,2, 3,35	6,22, 26,1	19,35, 29,13		19,1, 29	18,15, 1	15,10, 2,13		35,28	3,35, 15
36 Complexity of device	2,13, 28	10,4, 28,15		2,17, 13	24,17, 13	27,2, 29,28		20,19, 30,34	10,35, 13,2	35,10, 28,29		6,29	13,3, 27,10
37 Complexity of control	27,3, 15,28	19,29, 39,25	25,24, 6,35	3,27, 35,16	2,24, 26	35,38	19,35, 16	19,1, 16,10	35,3, 15,19	1,13, 10,24	35,33, 27,22	18,28, 32,9	3,27, 29,18
38 Level of automation	25,13	6,9		26,2, 19	8,32, 19	2,32, 13		28,2, 27	23,28	35,10, 18,5	35,33	24,28, 35,30	35,13
39 Productivity	29,28, 10,18	35,10, 2,18	20,10, 16,38	35,21, 28,10	26,17, 19,1	35,10, 38,19	1	35,20, 10	28,10, 29,35	28,10, 35,23	13,15, 23		35,38

Appendix E **185**

Feature to Change \ Undesired Result (Conflict)	27 Reliability	28 Accuracy of measurement	29 Accuracy of manufacturing	30 Harmful factors acting on object	31 Harmful side effects	32 Manufacturability	33 Convenience of use	34 Repairability	35 Adaptability	36 Complexity of device	37 Complexity of control	38 Level of automation	39 Productivity
21 Power	19,24, 26,31	32, 15, 2	32, 2	19,22, 31,2	2, 35, 18	26,10, 34	26,35, 10	35, 2, 10,34	19,17, 34	20,19, 30,34	19,35, 16	28,2, 17	28,35, 34
22 Waste of energy	11,10, 35	32		21,22, 35,2	21,35, 2,22		35,22, 1	2,19		7, 23	35, 3, 15,23	2	28,10, 29,35
23 Waste of substance	10,29, 39,35	16,34, 31,28	35,10, 24,31	33,22, 30,40	10,1, 34,29	15,34, 33	32,28, 2,24	2,35, 34,27	15,10, 2	35,10, 28,24	35,18, 10,13	35,10, 18	28,35, 10,23
24 Loss of information	10,28, 23			22,10, 1	10,21, 22	32	27,22				35,33	35	13, 23, 15
25 Waste of time	10,30, 4	24,34, 28,32	24,26, 28,18	35,18, 34	35,22, 18,39	35,28, 34,4	4,28, 10,34	32, 1, 10	35, 28	6, 29	18,28, 32,10	24,28, 35,30	
26 Amount of substance	18, 3, 28,40	13, 2, 28	33, 30	35,33, 29,31	3,35, 40,39	29, 1, 35,27	35,29, 25,10	2,32, 10,25	15, 3, 29	3,13, 27,10	3,27, 29,18	8, 35	13,29, 3,27
27 Reliability		32,3, 11,23	11,32, 1	27,35, 2,40	35, 2, 40,26		27,17, 40	1, 11	13,35, 8,24	13,35, 1	27,40, 28	11,13, 27	1,35, 29,38
28 Accuracy of measurement	5,11, 1,23			28,24, 22,26	3,33, 39,10	6,35, 25,18	1,13, 17,34	1,32, 13,11	13,35, 2	27,35, 10,34	26,24, 32,28	28, 2, 10,34	10,34, 28,32
29 Accuracy of manufacturing	11,32, 1			26,28, 10,36	4,17, 34,26		1,32, 35,23	25, 10		26, 2, 18		26,28, 18,23	10,18, 32,39
30 Harmful factors acting on object	27,24, 2,40	28,33, 23,26	26,28, 10,18			24,35, 2	2, 25, 28,39	35,10, 2	35,11, 22,31	22,19, 29,40	22,19, 29,40	33, 3, 34	22,35, 13,24
31 Harmful side effects	24,2, 40,39	3,33, 26	4,17, 34,26						19, 1, 31	2, 21, 27,1	2	22,35, 18,39
32 Manufacturability		1,35, 12,18		24,2			2, 5, 13,16	35, 1, 11,9	2, 13, 15	27,26, 1	6,28, 11,1	8, 28, 1	35, 1, 10,28
33 Convenience of use	17,27, 8,40	25,13, 2,34	1,32, 35,23	2,25, 28,39		2, 5, 12		12,26, 1,32	15,34, 1,16	32,26, 12,17		1,34, 12,3	15, 1, 28
34 Repairability	11,10, 1,16	10,2, 13	25,10	35,10, 2,16		1,35, 11,10	1,12, 26,15		7, 1, 4, 16	35,1, 13,11		34,35, 7,13	1, 32, 10
35 Adaptability	35,13, 8,24	35,5, 1,10		35,11, 32,31		1,13, 31	15,34, 1,16	1,16, 7,4		15,29, 37,28	1	27,34, 35	35,28, 6,37
36 Complexity of device	13,35, 1	2,26, 10,34	26,24, 32	22,19, 29,40	19,1	27,26, 1,13	27,9, 26,24	1,13	29,15, 28,37		15,10, 37,28	15, 1, 24	12,17, 28
37 Complexity of control	27,40, 28,8	26,24, 32,28		22,19, 29,28	2,21	5,28, 11,29	2,5	12,26	1,15	15,10, 37,28		34, 21	35, 18
38 Level of automation	11,27, 32	28,26, 10,34	28,26, 18,23	2,33	2	1,26, 13	1,12, 34,3	1,35, 13	27,4, 1,35	15,24, 10	34,27, 25		5,12, 35,26
39 Productivity	1,35, 10,38	1,10, 34,28	18,10, 32,1	22,35, 13,24	35,22, 18,39	35,28, 2,24	1,28, 7,19	1,32, 10,25	1,35, 28,37	12,17, 28,24	35,18, 27,2	5,12, 35,26	

Appendix F
Case Study Containment Ring Problem (Impeller Burst) Solved using TRIZ[39]

Abstract

An air compressor on a airplane consists of a fan, a fan shroud (to control the direction of the air stream), and an armor steel containment ring. The engineered system is designed to contain the fragments from a maximum speed fan impeller burst. The problem is that the ring is too heavy (Figure 1). TRIZ helped Allied Signal develop a lighter patentable design.

Introduction

The design team used the Situation Questionnaire, Problem Formulator and the TRIZ tools. The questionnaire established the problem area and the resources available. The Problem Formulator reformulated the problem into many little problems. Knowledge-based tools provided recommendations. A five-step process is presented.

Step 1. Identify and document the problem

The Innovation Situation Questionnaire asks several questions which begin the process of changing your methods for attacking a problem. A few

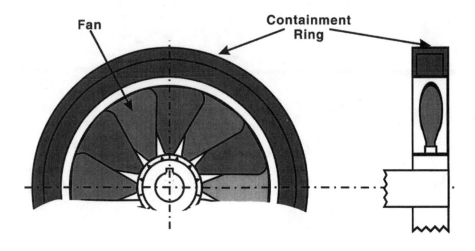

Figure 1. TRIZ help Allied Signal develop a lighter jet fan containment ring.

of the questions are presented on the next page. Looking at available resources as a design influence offers new insights into possibilities. Resource information should be organized in the QFD voice-of-the-customer table. A different purpose is now added to the customer site visit — identify resources to improve design options.

Answers to selected information requests follow:

Name the engineered system in which innovation is required, and the industry to which it belongs.

A containment ring is a part of a fan and belongs to the aircraft industry.

The technological system is designed to contain the fragments from a maximum speed fan impeller burst. The system consists of a fan, a fan shroud (to control the direction of the air stream) and an armor steel containment ring.

Identify the Primary Useful Function (PUF) performed or implemented by the system.

PUF: prevent the fragments produced by impeller burst from flying away.

Indicate the negative effect or drawback.

Drawback: the ring is too heavy.

After what events or developmental steps did the situation appear?

With the increasing size of airplanes, a more powerful fan was required. This more powerful fan had a higher speed of rotation and thus impeller fragments carried more energy. To contain the fragments a stronger ring was necessary, the ring was made thicker. The thicker ring weighs more.

Describe the mechanism, if it is understood, which causes the negative effect.

The containment ring must be strong to contain fragments. Fragments are produced by impeller burst. The impeller burst is caused by centrifugal forces which are the result of the high speed fan rotation.

Indicate the known or generally-used means or methods, if any, for eliminating the negative effect (drawback). Explain why you do not or prefer not to use these means or methods.

The general way to reduce weight is to reduce the amount of material. However, this leads to weakening of the ring, causing a safety hazard.

Describe what other problems could be solved if it is not possible to remove or eliminate the above indicated negative effect (drawback).

If there is no way to reduce the weight of the containment ring, one could try to prevent the impeller burst. However, this problem may be more difficult to solve than the initial one.

Describe your vision of the ideal result.

Ideally there is no containment ring at all, but all the necessary protection exists.

Step 2. Problem Formulation

The answers to the relationship questions in the two linked flow graphs (Chapter 3, Figures 1 and 2).

Create the problem formulation flow chart (Figure 2).

A partial list of Problem Statements which follow relate to each numbered node:

1. Find the way to prevent (Ring is heavy) under the condition of (Ring is thick).
2. Find an alternative way of (Ring is thick) which provides (High mechanical strength) and does not cause (Ring is heavy).

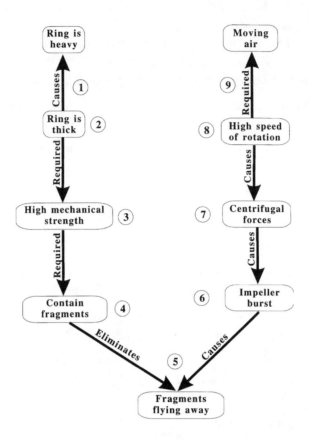

Figure 2. Problem formulation flow chart.

3. Find an alternative way of (High mechanical strength) which provides (Contain fragments) and which does not require (Ring is thick).
4. Find an alternative way of (Contain fragments) which prevents (Fragments flying away) and which does not require (High mechanical strength).
5. Find a way to prevent (Fragments flying away) which does not require (Contain fragments) under the condition of (Impeller burst).
6. Find a way to prevent (Impeller burst) under the condition of (Centrifugal forces).
7. Find a way to prevent (Centrifugal forces) under the condition of (High speed of rotation).
8. Find an alternative way of (High speed of rotation) which provides (Moving air) and does not cause (Centrifugal forces).

9. Find an alternative way of (Moving air) which does not require (High speed of rotation).

Step 3. Problem statements categorization

Analyze Problem Statements, identify Directions for Solution (innovation) and select the most promising solutions for further work.

Directions for Solution

Group 1. Redesign the containment ring.

- 1.1. Reduce the ring's weight without reducing mechanical strength. (Problem Statements 1, 2, 3)
- 1.2. Find alternative ways to absorb the fragments. (Problem Statements 4, 5)

Group 2. Work with other parts of the system.

- 2.1. Find ways to reinforce the fan to prevent the possibility of its bursting or to compensate for the destructive forces. (Problems Statements 6, 7)

Group 3. Problems are out-of-scope or too general (i.e., requiring critical changes to the existing system).

- 3.1. Develop a different process (rather than a rotating fan) for providing air. (Problem Statements 8, 9)

We will pursue Directions for Solution 1.1 and 1.2 from Group 1 above.

Step 4 and 5. Identifying and Using TRIZ Tools

We are going to use the System of Operators[40] to look for Solution ConceptsDirection 1.1. Reduce the ring's weight without reducing mechanical strength.

It is recommended that we consider two groups of Specialized Operators:

- Reducing weight
    - Abandoning symmetry
    - Reducing the weight of individual parts
    - Strengthening individual parts

- Increasing mechanical strength
    - Transforming an object's shape
    - Transforming an object's microstructure
    - Transforming the state of aggregation
    - Integration
    - Introducing a strengthening element
    - Anti-loading
    - Introducing an additive

Direction 1.2. Find alternative ways to absorb energy of the fragments.

It is recommended that we consider the following group of General Operators:

- Isolation
- Isolator — inexpensive substance
- Isolator — modification of available substances
- Self-isolation
- Introducing a liquid
- Selectively permeable isolation
- Using an easily-destroyed interlayer
- Using the culprit of an undesired action
- "Sheltering" for a period of time

We are going to consider the above Operators.
Ideas stimulated from the general operators:

1. Vary the thickness of the ring tube. Reduce the thickness where permissible.
2. Reduce the energy of the fragments by reducing their weight (i.e., help the impeller to break into smaller pieces). This will allow the ring to be made less strong and thus lighter.
3. Use a multilayer ring: additional strengthening rings, rings with different hardness and elasticity, rings which have a gap in between them, a gap filled with energy-absorbing material, etc.
4. Make a thin ring which has reinforcing ribs. If the ribs are placed on the internal surface of the ring, flying fragments will lose a large amount of their energy smashing into the ribs.
5. Find where the ring usually breaks and reinforce these places.

6. Introduce preliminary stress. For example, use additional rings which have been pressure-fitted to create a force directed toward the inside of the ring.
7. Use thermal treatment to harden the ring material.
8. Make the ring out of separate layers so cracks which develop inside won't spread.
9. Create inner stresses inside the ring. This can be done using wiring, banding, double-ring structures, etc.
10. Use special threads in the ring, such as those used in bullet-protective vests.
11. Use foam or foam-like filling to absorb energy shown in Figure 3.

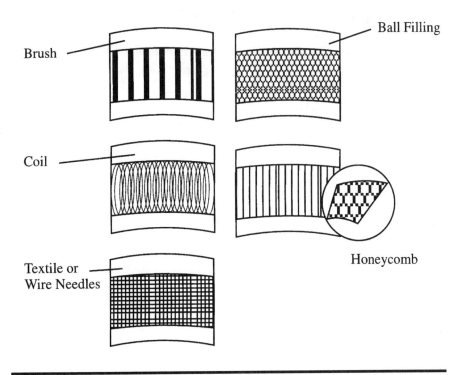

Figure 3. A filling absorbs energy.

Notes

1. It should be noted that most of the Solution Concepts which involve increasing the ring's mechanical strength may be applied to the fan itself; that is, by reinforcing the fan impeller to prevent it from bursting. For example, one can try to use reinforcing threads inside the blades, creating forces which oppose the destructive centrifugal forces. Additional ways of compensating for centrifugal forces can be obtained using the group of General Operators entitled "Counteraction."
2. Only a sample of the many ideas were listed above.

Conclusion

After thorough consideration, Subject Matter Experts decided to file a patent on a foam-like filling having an absorbing structure.

Appendix G
Resources[41]

Substance Resources

  Waste
  Raw materials and products
  System elements
  Inexpensive substance
  Substance flow
  Substance properties

Field Resources

  Energy in the system
  Energy from the environment
  Build upon possible energy platforms
  Systems waste becomes system energy

Space Resources

  Empty space
  Another dimension
  Vertical arrangement
  Nesting

Time Resources

  Pre-work
  Scheduling
  Parallel operations
  Post-work

Informational Resources
    Sent by substance
    Inherent properties
    Moving information
    Transient information
    Change of state information

Functional Resources
    Resource space within primary function
    Using harmful effects
    Using secondary generated functions

Glossary

AHP Analytic Hierarchy Process is a method of ranking items, events or options by using paired comparisons. The paired comparisons are structured to yield ratio data. Ratio data is necessary to make a ratio calculation in order to be able to state that "A" is twice as desirable a "B."

Analogous Thinking The mental process of drawing and applying the known solution of a resolved problem to the problem currently faced.

Anticipatory Failure Determination A method for systematically identifying and eliminating system failures — before they occur. This method, in effect, invents failure mechanisms and then examines the possibility of their actually occurring.

Solution Concept The fundamental and innovative technological idea that solves an inventive problem (that is, resolves the contradiction). Solution Concepts are nonnumeric in nature and require good engineering practice to become realizable solutions.

Contradiction The condition wherein contradictory requirements are placed upon a technological system. The terms paradox or oxymoron could also be applied to these situations.

Directed Product Evolution The systematic application of the Patterns of Evolution of Technological Systems to a current system with the intention of "forcing" its highly probable future development — before it occurs naturally. In effect, we systematically invent the future. Via this process, dominant market and patent positions can be systematically achieved.

Directed Product Improvement Improves a system by moving design in the direction of ideality.

Ideality Ideality is a qualitative assessment defined as the sum of a system's useful functions divided by the sum of its undesired characteristics (drawbacks). Ideal systems provide or implement all desired characteristics without any drawbacks. The use of existing resources along with physical, chemical, geometrical and other effects makes the ideal possible.

Inventive Problem   A problem which includes at least one contradiction for which a path to a solution is unknown.

Level of Innovation   TRIZ partitions problem solutions into five levels   Standard Solutions — Level 1, Improvements — Level 2, Innovations — Level 3, Inventions — Level 4, and Discoveries — Level 5.

Patterns of Evolution of Technological Systems   A compilation of trends within TRIZ which documents historically recurring strong tendencies in the development of manmade systems. These patterns can be applied to virtually any design, to understand its probable future and to accelerate realization of that future.

Physical contradiction   A contradiction wherein some element of a system is subject to two opposing requirements.

Resources   Resources are system/environment elements and their attributes, as well as sources of energy (heat, electricity, magnetism, motion, etc.) within and around (available to) the system that can be used in its movement towards the ideal.

Standard Solutions   The TRIZ knowledge base that has been developed includes: a compilation of contradictions and their solutions, at an abstract level ; also includes typical resources used in solution implementation that have been repetitively and successfully applied in the past.

Super Effect   An unplanned additional benefit resulting from improvement of the original problem.

Technical Contradiction   A contradiction wherein an improvement in one desired characteristic of a system results in the deterioration of another.

TRIZ   The Theory of Innovative Problem Solving is the result of nearly 50 years of research and development in the former Soviet Union by Genrich Altshuller and associates.

References

Preface

1. von Oech, R., *A Whack On The Side Of The Head*, Warner Book, New York, 1983.
2. Herrmann, N., *The Creative Brain*, Brain Books, Lake Lure, NC, 1988.
3. Ribaut, A., Russian translation from *Typography of Y. N. Erlich*, St. Petersburg, 1902.
4. Altshuller, G. S., lecture notes, 1970.

Chapter 1

5. Pugh, S., Total Design – Integrated Methods for Successful Product Engineering, Addison–Wesley, Reading MA, 1991.
6. Taguchi, G., *Introduction to Quality Engineering*, Asian Productivity Organization, Tokyo, 1983.
7. Altshuller, G. S., *Creativity as an Exact Science*, Gordon and Breach, New York, 17, 1988.
8. Gordon, W. J. J., *Synetics*, Harper & Row, NY, 1961.
9. Edison, T. A., said in 1903, in *Harper's Monthly Magazine*, 78, 6, September 1932.
10. Altshuller, G. S., Shapiro, P. B., "On the psychology of inventive creativity" (in Russian), *Voprosy Psikhologii (Questions in Psychology)*, 6, 37-49, 1956.

11. Shapiro, R. B., Expulsion of the seraph of creativity, State Publishing House of Children Literature of Russian Ministry of Culture (Detgiz) Moscow, 1961.
12. Braham, J., Inventive ideas grow on 'TRIZ', *Machine Design*, 60, October 12, 1995.
13. *Ideation Methodology: Course Notes*, Ideation International Inc., Southfield, MI, 1997.
14. Bar-El, Z., TRIZ methodology, *The Entrepreneur Network Newsletter*, May 1996.
15. Altov, H. (pen name of Altshuller, G. S.), translated by Lev Schulyak, *And Suddenly The Inventor Appeared*, Technical Innovation Center, Auburn, MA, 1994.
16. Osborne, A., *Applied Imagination*, Charles Scribner Sons, 1953.
17. Braham, J., Inventive ideas grow on 'TRIZ', *Machine Design*, 58, October 12, 1995.
18. Kaplan, S., *An Introduction to TRIZ: The Russian Theory of Inventive Problem Solving*, Ideation International Inc., Southfield, MI, 1996.

Chapter 2

19. *IWB Software*, Ideation International Inc., Southfield, MI, 1997.

Chapter 3

20. *The Formulator Software*, Ideation International Inc., Southfield, MI, 1996.
21. *IWB Software*, Ideation International Inc., Southfield, MI, 1997.
22. Zusman, A., Terninko, J., *TRIZ/Ideation Methodology for Customer-Driven Innovation*, 8th Symposium on Quality Function Deployment, The QFD Institute, Novi, MI, June 1996.

Chapter 4

23. *Ideation Methodology: The Training Manual (4th Ed.)*, Ideation International Inc., Southfield, MI, 169, 1995.

Chapter 5

24. Killander, A. J., *Generating Electricity for Families in Northern Sweden*, Report from The Department of Manufacturing Systems, Royal Institute of Technology, Stockholm, Sweden, 1996.
25. Gorin, Y., *Physical Effects and Phenomena for Inventors (Part 1)* (in Russian), Baku, Russia, 1973.
26. *Physical Effects for Inventors and Innovators, 2nd Ed.* (in Russian), Obninsk, Russia, 1977.
27. *Handbook of Chemistry and Physics, 73rd Ed.*, CRC Press, Boca Raton, FL, 1993.

Chapter 6

28. Ouspenski, P. D., *In Search of the Miraculous*, Harcourt, Brace and World Inc., New York, 1949.
29. *The Golden Age of TRIZ Software*, Ideation International Inc., Southfield, MI, 1996.

Chapter 8

30. Herrmann, N., *The Creative Brain*, Brain Books, Lake Lure, NC, 1988.
31. Terninko, J., *Step by Step QFD: Customer-Driven Product Design*, Responsible Management Inc., Nottingham, NH, 1995.
32. Terninko, J., *Robust Design: Key Points for World Class Quality*, Responsible Management Inc., Nottingham, NH, 1989.

Appendix A

33. Altshuller, G. S., Zlotin, B. L., Zusman, A. V., Filatov, V. I., *Searching for New Ideas From Insight to Methodology* (in Russian), Kartya Moldovenyaska, Kishnev, Moldova, 1989.
34. *Physical Effects for Inventors and Innovators, 2nd Ed.* (in Russian), Obninsk, Russia, 1977.

Appendix C

35. Kaplan, S., *An Introduction to TRIZ — The Russian Theory of Inventive Problem Solving*, Ideation International Inc., Southfield, MI, 1996.

Appendix D

36. *Ideation Methodology — The Training Manual (4th Ed.)*, Ideation International Inc., Southfield, MI, 1995.

Appendix E

37. *Ideation Methodology — The Training Manual (4th Ed.)*, Ideation International Inc., Southfield, MI, 1995.
38. *The Ideator Software*, Ideation International Inc., Southfield, MI, 1996.

Appendix F

39. Terninko, J., *Containment Ring Problem (Impeller Burst) Solved Using TRIZ*, 2nd Annual Total Product Development Symposium, American Suppliers Institute/University of California, Pomona, CA, November 1996.
40. *IWB Software*, Ideation International Inc., Southfield, MI, 1997.

Appendix G

41. *IWB Software*, Ideation International Inc., Southfield, MI, 1997.

Bibliography

Altov, H. (pen name of Altshuller, G. S.), translated by Lev Schulyak, *And Suddenly The Inventor Appeared*, Technical Innovation Center, Auburn, MA, 1994.
Altshuller, G. S., *Creativity as an Exact Science*, Gordon & Breach, New York, 1988.
Altshuller, G. S., Zlotin, B. L., Zusman, A. V., Filatov, V. I., *Searching for New Ideas: From Insight to Methodology* (in Russian), Kartya Moldovenyaska, Kishnev, Moldova, 1989.
Braham, J., "Inventive ideas grow on 'TRIZ'," *Machine Design*, 56-66, Oct. 12, 1995.
Bar-El, Z., "TRIZ Methodology," *The Entrepreneur Network Newsletter*, May 1996.
Gorin, Y., *Physical Effects and Phenomena for Inventors (Part 1)* (in Russian), Baku, Russia, 1973.
Handbook of Chemistry and Physics, 73rd Ed., CRC Press, Boca Raton, FL, 1993.
Herrmann, N., *The Creative Brain*, Brain Books, Lake Lure, NC, 1988.
Ideation Methodology: Course Notes, Ideation International Inc., Southfield, MI, 1997.
Ideation Methodology: The Training Manual, 4th Ed., Ideation International Inc., Southfield, MI, 1995.
Kaplan, S., *An Introduction to TRIZ: The Russian Theory of Inventive Problem Solving*, Ideation International Inc., Southfield, MI, 1996.
Killander, A. J., *Generating Electricity for Families in Northern Sweden*, "Report from The Department of Manufacturing Systems," Royal Institute of Technology, Stockholm, Sweden, 1996.
von Oech, R., *A Whack On The Side Of The Head*, Warner Book, New York, 1983.
Ouspenski, P. D., *In Search of the Miraculous*, Harcourt, Brace & World Inc., New York, 1949.
Physical Effects for Inventors and Innovators, 2nd Ed. (in Russian), Obninsk, Russia, 1977.
Pugh, S., *Total Design – Integrated Methods for Successful Product Engineering*, Addison-Wesley, Reading, MA, 1991.
Saaty, T., *Decision Making for Leaders*, University of Pittsburgh Press, Pittsburgh, PA, 1988.
Taguchi, G., *Introduction to Quality Engineering*, Asian Productivity Organization, Tokyo, 1983.
Terninko, J., *Systematic Innovation: Theory of Inventive Problem Solving (TRIZ/TIPS)*, Responsible Management Inc., Nottingham, NH, 1996.

Terninko, J., *Robust Design: Key Points for World Class Quality*, Responsible Management Inc., Nottingham, NH, 1989.
Terninko, J., *Step by Step QFD: Customer-Driven Product Design*, Responsible Management Inc., Nottingham, NH, 1995.
Terninko, J., *Introduction to TRIZ: A Workbook*, Responsible Management Inc., Nottingham, NH, 1996.
Terninko, J., *Containment Ring Problem (Impeller Burst) Solved Using TRIZ*, 2nd Annual Total Product Development Symposium, American Suppliers Institute/University of California, Pomona, CA, November 1996.
Zusman, A., Terninko, J., *TRIZ/Ideation Methodology for Customer-Driven Innovation*, 8th Symposium on Quality Function Deployment, The QFD Institute, Novi, MI, June 1996.

Software

The Golden Age of TRIZ, Ideation International Inc., Southfield, MI, 1996.
The Ideator, Ideation International Inc., Southfield, MI, 1996.
IM-Tech Optimizer, Invention Machine, Boston, MA, 1997.
IWB, Ideation International Inc., Southfield, MI, 1997.
Phenomenon, Invention Machine, Boston, MA, 1998.

Index

A

Acid, 95
Acid-filled container, 96
Adolescence, 131
Adult thinking, xi
Agriculture, 21
AHP, 132
Airplane, 80–81, 101, 103, 130
Altitude, 101
Analogy, 67
Analogic, 67
Analytic Hierarchy Process, 132
Analytic tool, 47, 91, 113
Altov 8–9
Altshuller, Genrich, iii, 4–5
Asymmetry, 71, 142
Author's certificate, 6
Automobile, 103, 133, 137
Auxiliary functions 98

B

Backpacking stove, 92
Balls, moving, 75
Ball-point pen, 138
Bathyscaph, 134
Bearing races, 102
Bearings, 107
Bicycle, 29, 32, 133
Bisystem, 137
Blood clot, 100
Brainstorming, 3
Bumpers, automobile, 133

C

Camp stove, 92
Cartier, Emie, ix
Cases
    Ford Motor Co., 10, 24
    Furnace, 51
    Rockwell Automotive, 10
Catamaran, 92
Cedar nuts, 22
Cement, 153
Chain, 135
Change operating principles, 103
Chemical welding, 100
Child thinking, xi
Childhood, 131
Clamp, 38
Clay pigeon, 110
Clots, 100
Coal dust explosions, 106
Conception, 128
Concrete, pre-stressed, 97
Consultant for hire, 19
Containment ring, 187
Contradiction
    Analysis, 65
    Physical, 65
    Table, 71
    Technical, 65, 67, 70
Cooling system, 112
Coring device, 38
Creative ideas, 25
Creative potential, 19
Customer, 152
Cyclical evolution, 111
Cyrillic words for TRIZ, 5

D

Decline, 132
Derived resources, 99
Diamond, splitting, 21
Discovery, 5

205

Dishes, 107
Duplication machines, 139
Dynamism, 133

E

Edison, Thomas A., 4
Effects, 97
Effective system, 114
Electrolytic process, 123
Electricity, 94
Energy engineers, 17, 74
Environment, 33
Evolution
    Product, 127
    Technology, 127
Evolution patterns, 127
    Decreased human involvement, 146
    Ideality, 132
    Increased complexity then simplification, 138
    Increased dynamism, 133
    Matching and mismatching, 141
    Toward microlevel and increased use of fields, 143
    Nonuniform development, 133
    S-curve, 128
Exhaust pipes, 105
External dynamization, 134

F

40 Principles, 15, 69
Field, 113
Field resources, 101
Filter cleaning, 22
Footprint, 144
Formulation process, 47, 149
Fruit tree, 140
Fuller, R. Buckminister, 112
Function, 47
Functional resource, 100
Fundamental problem, 19
Fundamental solution, 19
Furnace, 51

G

Generator, 94
Glass, 103–104
Gordon, William, 3
Growth, 131
Gyorgyi, Albert, Szent, ix

H

Hammer, 118
Harmful function, 47
Harmful system, 114, 119
Heat exchanger, 135
Herrmann, Ned, 147
Heuristics, 5
Honda, 66
Hot air balloon, 134
Hydrogenerator, 139

I

Ideal design, 91, 132
Ideality, 95, 132
Ideation International, Inc., 12, 214
Implementation, 147
Incomplete system, 114
Ineffective system, 114, 119
Increased dyamism, 133
Increased complexity, 138
Innovation, 2
Innovation process, 20
Innovative Situation Questionnaire, 29, 187
Internal dynamization, 134
Invention, 5
    Numbers, 128
Invention Machine Co., 12, 214
Inventive principles, 71
Inventive problem, 13, 70
Inventiveness, levels of, 7, 13, 15, 128
Iron, molten, 144
ISQ, 29, 148, 187

K

Kaplan, Stan, 26
Kishinev School, 29
Knowledge, 15–16
Knowledge based tools, 65, 127

L

Lasers, 14
Landing planes, 103
Landing-gear, 80–81
Level of inventiveness, 128
Limits, 133
Limber mill, 88

Index

M

Matching and mismatching, 141
Maturity, 132
Measure
   Temperature, 106
   Thickness, 106
Melting ore, 51
Macro to micro, 143
Mobile object, 137
Moldova, 29
Molten iron, 144
Motor mounts, 71, 72

N

Natural resource, 107
New England, 123
Nickel plating 81–82
Nonuniform development, 133
Number of inventions, 128

O

Osborne, A 19
Operator 191
Opportunity for improvement 35
Overheating 106

P

Painting, 98
Pappos, 5
Paradigm shift, 17
Patents, 21, 68
Patterns of evolution, 127
   Decreased human involvement, 146
   Ideality, 132
   Increased complexity then simplification, 137
   Increased dynamism, 133
   Matching and mismatching, 141
   Toward microlevel and increased use of fields, 143
   Nonuniform development, 133
   S-curve, 128
Pendulum, 102
Peppers, sweet, 21
Picking the tool, 150
Pipe cracks, 51, 112
Planning, 45
Plow, 110
Pollution, 105,
Polysystem, 138
Power station, 105
Pregnancy, 128
Premises, 3
Primary useful function, 30
Problem, 35
Problem definition, levels, 55
Problem formulation, 53, 57
   Screw, 58, 61
   Furnace, 53
Profitability, 128
Psychological inertia, 17
PUF, 47
PHF, 47

Q

QFD, 1,147
Quotes
   Bacon, Francis, 45
   Ellis, Havelock, 27
   Fuller, R. Buckmister, 112
   Nichol, F. W, 63
   Steffens, Lincoln, 89
   Thoreau, Henry, David, 125
   Wells, H. G., 146
   Wordsworth, William

R

Replace, 102
Resource, 34, 91, 93, 99, 104
   Derived, 99
   Field, 101
   Functional, 100
   Substance modification, 99
   Waste, 99, 105, 107
Ribaut, Antwan, ix
Robust Design, 152
Rock, 117, 123
Rollers, 103
Root cause, 36
Rope, 135
Rosoy, 3
Russian acronym, 5

S

S-curve, 128
Scientist, 70
Screw, 29, 32
   Mechanical vibration, 73
   Resonant frequency, 74
Secondary problem, 39, 78

Seebeck effect, 93
Seeds, 140
Self-service, 101
Separation principles, 81
   In space, 81
   In time, 83
   Upon condition, 87
   Within a whole object, 84
Shape memory metals, 14, 136
Simplification, 138, 103
Simulating landing, 103
Skeet shooting, 110
Slag, 86
Snow, 106
Soap, 107
Soldering iron, 86
Solution space, 66
Sodium bicarbonate, 108
Spoilers, 137
Stalin, Joseph, 8
Stove, 92
Stapler, 139
Su–Field, 113
Substance modification, 99, 107, 109
Substance resources, 99
Substance–Field, 113
Substances, 117
Sugar, powdered, 22
Sunflower seeds, 22
Super-system, 40
Surface area, 136
Standard solution, 68, 124
Stone, splitting, 123
Stake holder, 152
Sweden, 94
Synergy, 1, 151
Synetics, 3
Systematic innovation, 2–3

T

39 Parameters, 68
Taguchi, Genichi, 1, 152

Tea pot, 106
Team composition, 148
Technical contradiction, 70
Technologies, major, 15
Temperature, measure, 106
Thermal memory, 14
Thickness, measure, 106
Time resources 111
Transferable solutions, 15
Transistors, 14
Triangle, 114
TRIZ thinking, xi
TRIZ, 1
TRIZ School, 11
Turning heavy
          rotors, 102

U

Use of resources, 97
Useful function, 47
Utilize resource, 104

V

Vacuum pump, 55

W

Waste resources, 99, 109
Washing
   Dishes, 109
   Electrolyte, 123
White-out tape, 138
Wide screen, 83
Wire drawing, 83
Wright brothers, 131

X

X-ray, 31
Washing machines, 146
Weak link, 133